Laboratory Manual of Fluid Mechanics and Machines

Laboratory Manual of Fluid Mechanics and Machines

V.P. Gupta
Associate Professor of Civil Engineering

Jagdish Chandra
Professor of Civil Engineering

K.S. Gupta
Professor of Mechanical Engineering

Faculty of Engineering
University of Jodhpur
Jodhpur (Raj.)

CBS

CBS Publishers & Distributors Pvt. Ltd.
NEW DELHI • BANGALORE • PUNE • COCHIN • CHENNAI (INDIA)

ISBN : 81-239-0009-0

First Reprint : 1992
Reprint : 1999, 2002, 2006,
2009, 2010, 2012

Published by Satish Kumar Jain and produced by V.K. Jain for
CBS Publishers & Distributors Pvt. Ltd.,
CBS Plaza, 4819/XI Prahlad Street, 24 Ansari Road, Daryaganj,
New Delhi - 110002, India. • Website: www.cbspd.com
e-mail: delhi@cbspd.com, cbspubs@airtelmail.in
Ph.: 23289259, 23266861, 23266867 • Fax: 011-23243014

Branches:

- ***Bengaluru:*** Seema House, 2975, 17th Cross, K.R. Road, Bansankari 2nd Stage, Bengaluru - 560070 Ph.: +91-80-26771678/79 • E-mail: cbsbng@gmail.com, bangalore@cbspd.com
- ***Pune:*** Bhuruk Prestige, Sr. No. 52/12/2+1+3/2, Narhe, Haveli (Near Katraj-Dehu Road By-pass), Pune - 411051 Ph.: +91-20-64704058/59, 32342277 • E-mail: pune@cbspd.com
- ***Kochi:*** 36/14, Kalluvilakam, Lissie Hospital Road, Kochi - 682018, Kerala • Ph.: +91-484-4059061-65 Fax: +91-484-4059065 • E-mail: cochin@cbspd.com
- ***Chennai:*** 20, West Park Road, Shenoy Nagar, Chennai - 600030 Ph.: +91-44-26260666, 26208620 E-mail: chennai@cbspd.com

Printed at :
J.S. Offset Printers, Delhi

पूज्यनीय जियाजी

के

चरण कमलों

में

समर्पित

PREFACE

(Third Edition)

Encouraged by the appreciative response this *Manual of Fluid Mechanics and Machines* has been consistently receiving from the students and teachers, the authors are pleased to bring out the third edition of the book.

This edition has been revised and enlarged by including new experiments. This will help to understand better the theory of transition from laminar to turbulent flow, verification of momentum theorem, Stokes law and elbowmeter. It is believed that the enlarged and revised edition of this Manual will find a still greater favour among the students and will be more useful to them.

Further suggestions for the improvement of the Manual will be most appreciated and acknowledged.

JODHPUR,
September, 1989

V.P. Gupta
J. Chandra
K.S. Gupta

PREFACE

(Second Edition)

Laboratory Manual of Fluid Mechanics and Machines is the revised version of the authors' earlier book (1985) 'Manual of Fluid Mechanics and Machines, Vol. 1—Laboratory Experiments' which covers the laboratory experiments on fluid flow more commonly required for the Civil and Mechanical Engineering degree and diploma courses. Starting with the first introductory part on definitions, flow variable measurements and basic equations, there are eleven experiments on Conduit Flow in the second part, seven experiments on Open Channel Flow in the third part and five experiments on Hydraulic Machines in the fourth part of the book The experiments are presented in the standard format of Object, Theory, Experimental set up Procedure, Calculations, Presentation of result, Precautions and viva-voce. Standard results/graphs are given as a guide. There are blank observation tables and necessary graphs for completion. With the clarity of presentation the students may be able to perform the experiments by themselves even without detailed instructions.

The authors are thankful to Professor Alam Singh who has always been a source of inspiration in the preparation of the text.

The authors expect the Manual to be very useful to the students for performing the experiments with understanding and interest. The teachers may appreciate it for planning experimentation and making their instructional part easy.

V.P. Gupta
J. Chandra
K.S. Gupta

Jodhpur
Sankranti, 14 Jan. 1988

CONTENTS

PART 1

Introduction

1.1 DEFINITIONS

1 FLUID MECHANICS It is that division of engineering which deals with the behaviour of fluid when subjected to a system of forces.

2 FLUID It is defined as the substance which deforms or yields continuously due to application of shear stress, no matter how small it is. It comprises two states of matter, i.e. liquid and gas.

3 IDEAL FLUID An ideal fluid may be defined as the one which is non viscous and incompressible.

4 REAL FLUID All the fluids have got viscosity to some extent and can be compressed to some extent and are known as real fluids.

5 SPECIFIC MASS (ρ) It is the amount of fluid per unit volume. The unit of specific mass is metre slug/m^3. (MKS) or kg/m^3 (SI). Mass density of water at sea level at a temperature of 4°C is 102 m slug/m^3 (MKS) or 1000 kg/m^3 (SI).

6 SPECIFIC WEIGHT (w) It is defined as the gravitational pull acting on unit mass falling freely and its unit is kN/m^3 (SI)

$$w = \rho g$$

for water at mean sea level at 4°C, the specific weight is 9810 kN/m^3.

7 VISCOSITY (μ) It is the property of fluid by virtue of which it offers resistance to shear or angular deformation. All those fluids which follow the Newton's law of viscosity are known as Newtonian fluids. ($\tau = \mu\ \partial u/\partial y$); where $\partial u/\partial y$ is the angular deformation caused to the fluid mass due to the application of shear stress τ kN/m^2. The constant of proportionality is known as dynamic viscosity. The unit of μ is Newton second/metre^2. Generally, it is expressed in poise (=dyne sec/cm^2). Viscosity of water is 1 centipoise (=1/100 poise)

$$1 \text{ poise} = 0.1 \text{ N s/m}^2$$

8 KINEMATIC VISCOSITY (ν) The ratio of dynamic viscosity to mass density ρ is known as kinematic viscosity ($=\mu/\rho$). The unit of measurement of ν is m^2/s. Generally, it is measured in stokes (=1 cm^2/sec).

9 PRESSURE (p) It is defined as the force per unit area

$$p = \lim_{\Delta A \to 0} \frac{\Delta F}{\Delta A}$$

It is measured in pascal (=N/m^2) and it acts normal to the surface. The hydrostatic pressure due to head h is given by

$$p = wh$$

10 DISCHARGE (Q) It is the volumetric rate of fluid flow and is measured in m^3/s.

$$Q = \text{Volume/Time} \quad m^3/s$$

1.2 MEASUREMENT OF FLOW QUANTITIES

1 DIFFERENTIAL PRESSURE It is measured by means of a U-tube manometer. For measuring large difference of pressure, U-tube mercury manometer is used. In case of small difference of pressure, inverted U-tube manometer is used and a lighter liquid is used in the manometer.

The difference of pressure Δh is given by

U-tube: $(p_1/ws+z_1) - (p_2/ws+z_2) = \Delta h = x\,(S_m/S - 1)$

Inverted U-tube: $\Delta h = x\,(1 - S_m/S)$

2 DISCHARGE MEASUREMENT Generally volumetric method is used to measure the discharge. The water is collected in a tank of area A (in plan). If the water level aries from H_1 to H_2 in time t seconds,

$$Q = A\,(H_1 - H_2)/t \quad m^3/s$$

Additionally, various shaped notches, (rectangular, triangular, proportional etc.) are also used to measure the discharge.

Rectangular notch: $Q = \frac{2}{3}\, Cd \sqrt{2g}\; L\, H^{3/2}$

Triangular notch: $Q = (8/15)\, Cd \sqrt{2g}\; \tan \theta/2\; H^{5/2}$

Proportional notch: $Q = C_1 H + C_2$

where C_1 and C_2 are numerical constants and θ is the apex angle of V-notch.

3 REYNOLDS NUMBER It is a dimensionless number which defines the influence of viscous forces over inertia forces. It is denoted by R_e and is generally used for enclosed flows.

$$R_e = \rho\, U\, D/\mu$$

4 FROUDE'S NUMBER It is a dimensionless number which defines the influence of gravity forces over inertia forces and is denoted by N_F and is generally used for free surface flows.

$$N_F = U/\sqrt{gD}$$

1.3 BASIC EQUATIONS

The basic equations used in fluid mechanics are

(1) Conservation of mass - continuity equation.
(2) Conservation of energy - energy equation.
(3) Conservation of momentum - momentum equation.

1 CONTINUITY EQUATION If a fluid flows with a velocity U_1 at section 1 (cross sectional area A_1) to section 2 (area A_2) where its velocity becomes U_2,

$$A_1 U_1 = A_2 U_2$$

2 ENERGY EQUATION The principle of conservation of energy states that neither the energy can be created nor it can be destroyed.

The fluid mass possesses potential energy by virtue of its elevation (datum energy), kinetic energy due to its velocity (U) and pressure energy. Generally, these energies are expressed in terms of equivalent height of liquid, i.e. head, hence are measured in the length units. (Fig. 1.1)

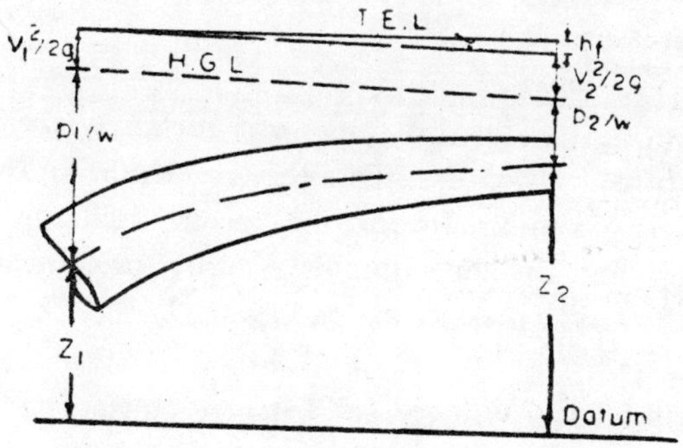

Fig. 1.1 Flow through a stream tube

Energy at section 1 = Energy at section 2.

$$\text{or } \frac{p_1}{w} + z_1 + \frac{U_1^2}{2g} = \frac{p_2}{w} + z_2 + \frac{U_2^2}{2g}$$

for an ideal, incompressible fluid. It is known as Bernoulli's theorem. But, in practice no ideal fluid is found and all the real fluids possess viscosity, however small it is. This, in turn, will cause certain loss of energy h_L from sections 1 and 2.

$$H = \frac{p_1}{w} + z_1 + \frac{U_1^2}{2g} = \frac{p_2}{w} + z_2 + \frac{U_2^2}{2g} + h_L$$

The term $\frac{p_1}{w} + z_1$ is known as piezometric head and the summation of all three terms is known as total head. If some machine is interposed (e.g., a pump) between sections 1 and 2 and supply energy E_m to the fluid, then

$$H_1 + E_m = H_2$$

In case it is taking out some energy (e.g., a turbine), then

$$H_1 - E_m = H_2$$

3 MOMENTUM EQUATION The summation of all the externally applied forces on a given control volume is equal to the rate of change of momentum brought in the given control volume in the direction of forces.

$$\Sigma F_x = (\rho_2 Q_2 U_{2x} - \rho_1 Q_1 U_{1x})$$
$$\text{and } Q_1 = A_1 U_1; \; Q_2 = A_2 U_2$$

The resultant force acting on the boundary, e.g. pipe or vane will be equal in magnitude and opposite in direction to that calculated above.

4 OPEN CHANNEL FLOW Flows occurring under the influence of gravity are termed as open channel flow. These have a free surface (i.e., the pressure is atmospheric on the liquid surface).

5 SPECIFIC ENERGY The total energy at any section of an open channel measured with reference to its bed is known as specific energy.

$$E = y + \frac{U^2}{2g} = y + \frac{Q^2}{2g\,A^2}$$

For given discharge Q, it yields two values of y for any value of E. These are known as alternate depths. Of course, there exists a depth, known as critical depth 'y_c when the specific energy is minimum for the given discharge Q. Additionally, the discharge is maximum and Froude's number is unity for critical flow.

This condition of flow is expressed as

$$Q^2T/gA^3 = 1$$

where T is the top width of the cross section at the free surface.

Specific force

The application of momentum equation leads to specific forces being equal at the inlet and outlet sections of the given control volume.

$$\text{Specific force } F_s = A\,\bar{x} + \frac{Q^2}{gA}$$

For critical flow, the specific force is also found to be minimum.

Simplifying and solving further for the case of rapidly varied flow (hydraulic jump) leads to

$$\frac{y_2'}{y_1'} = \tfrac{1}{2}\left[-1 \pm \sqrt{1-8\,N_{F_1}^2}\,\right]$$

where y_1' is the initial depth, y_2' is sequent to y_1'. These two depths are known as the conjugate depths (i.e. for the given specific force, two possible depths are obtained.)

Hydraulic machines

1 SPECIFIC SPEED (TURBINE) The specific speed of any turbine is the speed in rpm of a turbine geometrically similar to the actual turbine, but of such a size that under corresponding conditions it will develop 1 metric horse power when working under unit head (i.e. 1 metre).

$$N_s = N\sqrt{P}/H^{5/4}$$

N_s varies from 20 to 35 for Pelton wheel, 60 to 300 for Francis turbine and 300 to 1000 for Kaplan turbine.

2 SPECIFIC SPEED (CENTRIFUGAL PUMP) It may be defined as the speed in rpm of a geometrically similar pump of such a size that under corresponding conditions it would deliver 1 litre of liquid per second against a head of 1 metre

$$N_s = N\sqrt{Q}/H^{3/4}$$

It is to be noted that Q is used in litres/sec whereas H is used in metres.

The specific speed of the hydraulic machines is determined corresponding to maximum efficiency. For this purpose firstly the performance curves (characteristic curves) are obtained in the laboratory. In case of turbines, the required values of 'P' and 'N' corresponding to maximum efficiency is read from the curves. Similarly, the required values of Q and H are read from operating characteristics of the given centrifugal pump.

PART 2

Experiments on Conduit Flow

2. Place the Prandtl's tube at the tunnel bottom and take the manometer reading.
3. Raise the Prandtl's tube by 2 mm and repeat step 2.
4. Repeat the steps till centre of wind tunnel is reached or no change in the manometer reading is noted.

Note: Use inclined U-tube manometers or electronic instruments (pressure transducers) for pressure measurements.

Calculations

1. Calculate the velocity of the fluid at any distance y, from tunnel boundary.
   $u = C\sqrt{2g\,\Delta h}$ (col. 5)
2. Plot the velocity u versus y and determine the free stream velocity U_∞.
3. Determine the distance y where $u = 0.99\,U_\infty$ which is nominal B.L. thickness.
4. Find the area of the shaded portion of the velocity defect profile (Fig. 2).
   $\delta_* =$ Shaded area/U_∞
5. Plot u/U_∞ vs. y. Divide this plot into small strips and determine the product $(u/U_\infty)(1 - u/U_\infty)\,dy$.
6. Add all such products
   $\theta = \int (u/U_\infty)(1 - u/U_\infty)\,dy$
7. Plot u/U_∞ vs. y on log-log graph paper. Fit in a straight line and determine its slope.
8. Plot θ verses X.
9. The wall shear is determined from Eq. 3.

Presentation of results

Nominal boundary layer thickness $\delta =$
Displacement boundary layer thickness $\delta_* =$
Momentum boundary layer thickness $\theta =$
Shear stress $\tau_0 =$
Exponent 'n' = slope of log u/U_∞ vs. log y/δ plot =

Precautions

1. Read the inclined manometer correctly.
2. Draw the smooth curve between θ and X.

Viva-voce

1. Define boundary layer thicknesses δ, δ_* and θ.
2. What will be the effect of pressure gradient on the growth of boundary layer?

2.1 BERNOULLI'S THEOREM

Object

To verify the Bernoulli's theorem experimentally.

Theory

The principle of conservation of energy gives rise to Bernoulli's theorem which states that if an incompressible ideal fluid (i.e. no losses and specific mass constant throughout) flows through a closed passage, the total head, i.e. the sum of datum head z, pressure head p/w and velocity head $U^2/2g$ will be constant at all points, i.e.

$$H = p/w + z + U^2/2g = \text{constant at all points.}$$

But the validity of above theorem is subject to (i) the flow being steady, (ii) fluid being frictionless, that is, nonviscous (iii) flow being irrotational and (iv) no external work is done on the flow system by any external machine.

Experimental set up

The apparatus consists of a converging diverging duct (e.g. 38 mm width) issuing out of a cylindrical water container and discharging into another. Piezometric connections are made at small equal intervals to show pressure heads at different sections. The discharge is measured by collecting a known volume of water in time t. The water is supplied to the apparatus from the laboratory storage tank which can be regulated by inlet valve. Another valve is fitted at the end of duct to get sufficiently low pressure at the central tube. A graph is attached on a vertical board to take the piezometric readings. Alternatively, a scale may be used for the same purpose.

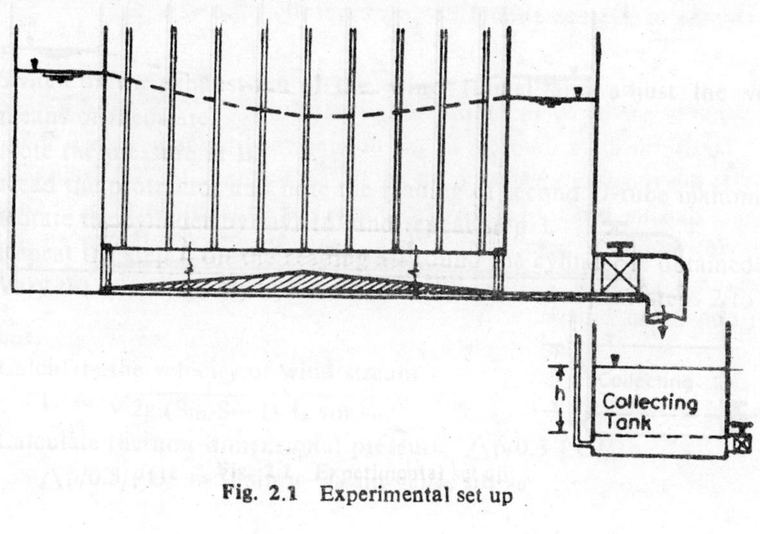

Fig. 2.1 Experimental set up

Procedure

1. Open the inlet valve to obtain a steady flow. Collect the water in a tank for certain time t. (col. 2, 3).

2. Consider the base of the appartus as base datum; measure the height of water level in the tubes above base, $p/w+z$ (col. 9).

Note: p/w is the pressure head and z is the datum head and the sum of these two quantities is called piezometric head. By proper adjustment of outlet valve, sufficiently deep curve can be obtained.

(3) Note down the width, b, and depth, d, of the duct at various cross sectionsl (col. 5).

Calculations

1. Calculate the area of cross section 'a' at a given section $\Rightarrow bd$ (col. 6)
2. Discharge $Q = A\ (h_2 - h_1)/t$ (col. 4)
3. Velocity of flow $U = Q/a =$ (col. 4) / col. 6.
4. Velocity head $= U^2/2g$

Presentation of results

1. Mark on a graph the ordinates of $(p/w+z)$ on a suitably chosen horizoutal scale which is called hydraulic gradient line (i.e. H.G.L).

2. Mark velocity head $U^2/2g$ over $(p/w+z)$ curve. It is called total energy line i.e. T.E.L.

3. Show that the resulting curve is a mildly sloping straight line, adding to it the loss of head, results in a horizontal line. Thus, it proves the Bernoulli's theorem

Precautions

(1) Discharge should be kept constant throughout one set of observations.
(2) Depths of passage should be very carefully observed.

Viva-voce

(1) What is meant by incompressible ideal fluid ? (2) What difference does it make to the Bernoulli's theorem if the flowing medium is not ideal fluid but a rea one ? (3) Is the theorem applicable to gases and vapours too ? Discuss this point ? (4) Are these datum, pressure and velocity energies interconvertible ? Give example of each ? (5) What is meant by irrotational flow ? (6) If a prime mover or any other machine is installed between any two flow sections, how is it accounted for in Bernoulli's equation ?

Observations

Width of passage b =

Area of collecting tank A =

Set I

Sl. No.	Discharge			Depth of passage	Area of passage	U	$U^2/2g$	$p/w+z$	H	Loss of head
	Rise of water level (h_2-h_1)	time t	Q							
1	2	3	4	5	6	7	8	9	10	11

Set II

Sl. No.	Discharge			Depth of passage	Area of passage	U	$U^2/2g$	$p/w+z$	H	Loss of head
	Rise of water level (h_2-h_1)	time t	Q							
1	2	3	4	5	6	7	8	9	10	11

2.2 VENTURIMETER

Object

1. To calibrate the Venturimeter.
2 To obtain a graph between discharge coefficient 'K' and $\log_{10}R_e$

Theory

By applying Bernoulli's theorem between entrance and throat sections, the relation for discharge can be derived.

$$Q = K \frac{A\sqrt{2g}\ \sqrt{\Delta h}}{\sqrt{(D/d)^4 - 1}} = K.C.\sqrt{\Delta h}$$

Here Δh is the difference of piezometric heads between entrance and throat sections, and is known as "venturi head." C is called constant of venturimeter and depends upon geometric parameters of given venturimeter and K is called discharge coefficient. It takes into account the energy loss between two sections and depends on the Reynolds number of flow Re (Fig. 2.2.1). Usually K varies from 0.96 to 0.99 for Re$>10^5$. The Reynolds number Re is given as

$$U_2\, d/\nu = 4Q/\pi d\nu$$

ν is the kinematic viscosity of fluid and depends on the temperature.

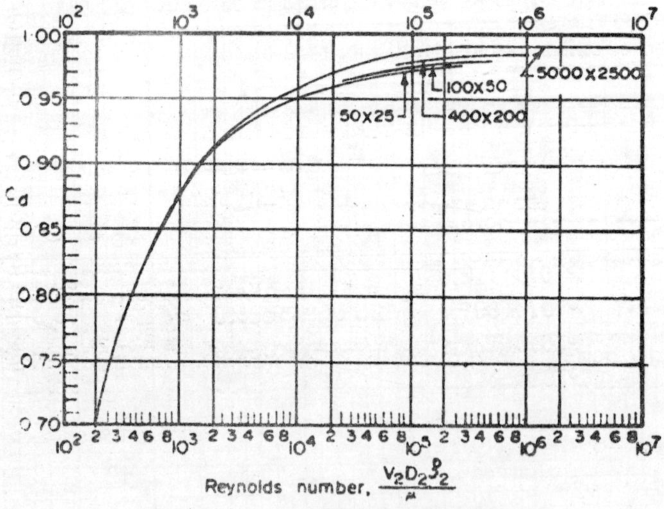

Fig. 2.2.1 Variation of K with log Re

Experimental set up

The Venturimeter is fitted in a pipe line, to which an inlet valve is fitted. The pressure taps are provided at entrance section 1 and throat section 2. A U-tube mercury manometer is fitted between these points 1 and 2. Water tank and a stop watch is used to measure the discharge. (Fig. 2.2.2)

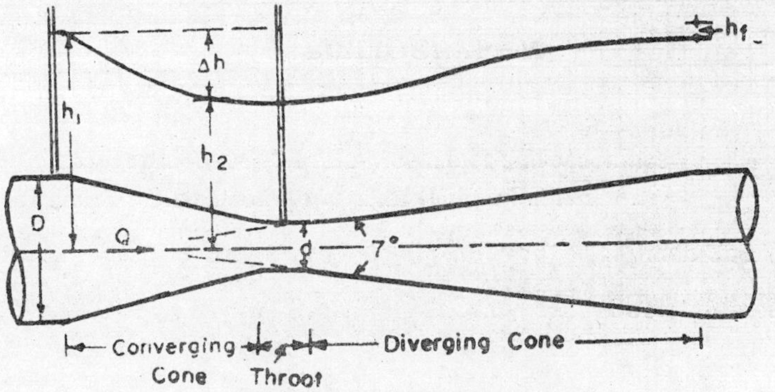

Fig. 2.2.2 Experimental set up

Procedure

1. Regulate the inlet valve to allow a steady flow through venturimeter.
2. Collect the certain volume of water (depth varing from h_1 to h_2) in the tank and note the time t.
3. Note the manometer readings in the left limb (h_1) and right limb of manometer (h_2). This difference is x.
4. Repeat the steps 1 to 3 for other rate of flow.
5. Take seven to eight sets of readings.

Calculations

1. The venturimeter constant C is calculated as

$$C = A\sqrt{2g} / \sqrt{(D/d)^4 - 1}$$

2. The discharge Q is then calculated by volumetric method (col. 5)
3. The deflection of mercury manometer x (col. 6) is converted into equivalent head of flowing liquid (water)

$$h = x\,(S_m/S - 1) = x\,(13.6/1 - 1) = 12.6\,x \quad \text{(col. 7)}$$

where S_m is the specific gravity of measuring fluid and S is the specific gravity of water.

4. Now the square root of h is determined.
5. The discharge coefficient K is calculated.

$$K = Q / C\sqrt{h} \qquad \text{(col. 9)}$$

6. Finally the Reynolds number R_e (col. 10) is calculated corresponding to different set of discharges.

Presentation of result

(1) Plot a graph between Q and $\sqrt{h}$ on an ordinary graph. It will be a straight line passing through the origin (Graph 1).

(2) Plot another graph between K and R_e on a semi-logarthmic graph sheet (Graph 2). Average value of K =..............

Observations

Diameter of pipe D = cm

Diameter of throat d = cm

Laboratory temperature = °C

Kinematic viscosity = m^2/sec

Specific gravity of manometer liquid = 13.6 (mercury)

Area of collecting tank A = m^2

Sl. No.	Discharge reading			$Q=\frac{A_1 \Delta H}{t}$ m^3/sec	Venturimeter deflection 'X' cm	h cm	$h^{1/2}$	K	Re	log Re
	Initial m	Final m	Time sec							
1	2	3	4	5	6	7	8	9	10	11

Precautions

1. Drive out all entrapped air from differential mercury manometer.
2. Maintain a constant discharge before taking any reading.

Viva-voce

1. What is the principle and use of venturimeter ? Can it be used for large penstokes. How will pressure corrections be made then ?

2. Why is the length of converging cone smaller than the diverging cone of venturimeter ?

3. Why is the pressure difference between entrance and throat section increased due to friction ?

4. If the direction of flow is reversed what will be the effect on the value of 'h', (i) if friction is neglected. (ii) friction is taken into account ?

5. Accuracy of result depends on how great is the value of h. If we reduce 'd' further, the value of h increases. Can we go on reducing throat diameter 'd' as much as we please ? If not what is the factor to limit the value of D/d ?

6. Just by seeing a separate venturimeter, can you ascertain the direction of flow, how ?

2.3 ORIFICEMETER

Object

1. To calibrate the given orificemeter.
2. To study the variation of the coefficient of discharge with the Reynolds number for two different values of d/D.

Theory

By applying Bernoulli's theorem between sections 1 and 2 (see Fig. 2.3 2) and using continuity equations, the discharge is given by

$$Q = Cd\, a\sqrt{2\, g.\, \Delta h}$$

Here Q is the discharge flowing in pipe, 'a' area of orifice contained in the orifice plate, Δ h is difference of pressure heads between pipe section upstream and the vena contracta section behind the orifice and Cd is coefficient of discharge of the orificemeter.

To study the influence of fluid kinematic viscosity on the value of Cd, the latter would be graphically plotted against log R_e; R_e being the Reynolds number at the orifice section (Fig. 2.3 1).

$$R_e = \rho U_2 d/v = 4Q/\pi d v$$

where d is diameter of orifice, v is kinematic viscosity of fluid at the laboratory temperature.

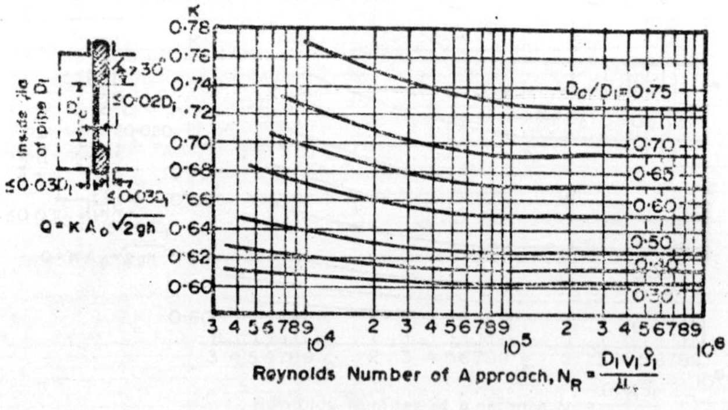

Fig. 2.3.1 Variation of Cd with Re

Experimental set up

The set up (Fig. 2.3.2) consists of a long horizontal pipe line. A thin plate having a concentric sharp edged circular orifice of diameter 'd' is fitted in the pipe line. Sufficient straight length is provided on the upstream of the orifice plate.

A valve is fitted at the end of pipe to regulate the discharge Q. The pressure tappings, one on the upstream side at (D-d) to 2 (D-d) and the other on the downstream side at 0.5 are provided on the horizontal pipe line. A U-tube mercury manometer is used to measure the pressure difference between sections 1 and 2. A stop watch is also required to measure time.

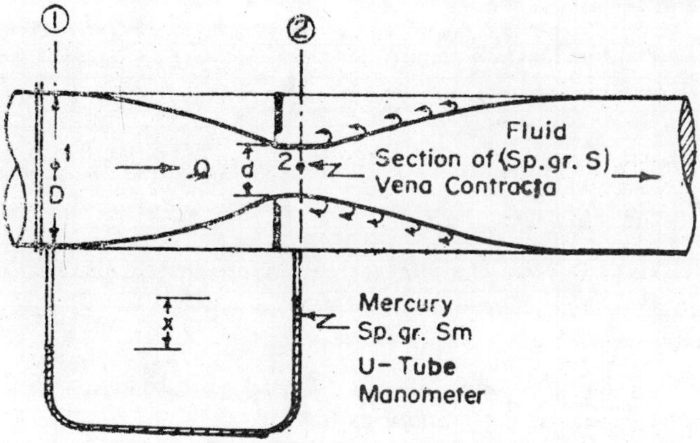

Fig. 2.3.2 Experimental set up

Procedure

1. Regulate the inlet valve to allow a steady discharge to flow through orificemeter.
2. Collect the water to a certain depth H in the collecting tank within time interval 't'.
3. Note down the deflection 'X' of mercury columns in the U-tube manometer.
4. Readjust the inlet valve for other rate of flow.
5. Repeat steps 1 to 3.
6. Take eight sets of readings.

Calculations

1. The discharge Q is calculated by volumetric method (col. 5)
2. The deflection of mercury manometer X (col. 6) is converted into equivalent head of liquid of specific gravity S flowing in pipe.

$$\Delta h = x\,(S_m/S - 1) = 12.6\,X \qquad \text{(col. 7)}$$

3. The square root of Δh is then computed in column 8.
4. The orificemeter constant is computed as

$$C = \pi/4\; d^2 \sqrt{2g}$$

5. The coefficient of discharge is calculated for different sets.

$$C_d = Q/C\sqrt{\Delta h}$$

6. Finally the Reynolds number Re (col. 10) is calculated corresponding to different sets of discharge.

Observations

Diameter of the pipe D = cm

Diameter of orifice in the orifice plate d= cm

Laboratory temperature = °C

Kinematic viscosity = m^2/s

Area of collecting tank A =

Specific gravity of manometer liquid = 13.6 (mercury)

Sl. No.	Discharge reading Initial m	Final m	Time sec	$Q = \frac{\Delta H}{t}$	Mercury deflection X m	Equivalent water head Δh m	$\sqrt{\Delta h}$	Cd	Re
1	2	3	4	5	6	7	8	9	10

Average value of Cd= ....

Presentation of results

1. Plot a graph between Q and $\sqrt{\Delta h}$ on an ordinary graph. The slope of this line is the coefficient of discharge multiplied by orificemeter constant C.

2. Prepare another graph between C_d and log Re on an ordinary graph sheet. Alternatively, use a semi logarithmic paper plotting C_d on ordinary scale and Re on logarithmic scale. Indicate the ratio d/D with this plot (Refer to Fig. 2.3.1).

Precautions

1. Remove all entrapped air from two limbs of manometer.
2. Maintain constant discharge for one set.
3. Take a number of readings at lower value of Reynolds number, i.e. at lower discharges.

Viva-voce

1. What is the use of an orificemeter ? Compare the advantages and disadvantages of an orificemeter and a venturimeter ?
2. How will you ascertain the direction of flow in an installed orificemeter with its pressure connections projecting out ?
3. Is there any factor to limit the minimum value of d/D that is workable ?
4. Why is the coefficient of discharge for orificemeter much less than that of a venturimeter, though both work on the same principle ?
5. It is required to keep a fairly long length of pipe free from bends, valves, or any other obstruction inside before the orifice section. What is the idea ?
6. What is the influence of factor d/D on the value of C_d ?
7. Conclude the effects of viscosity on the value of C_d ?

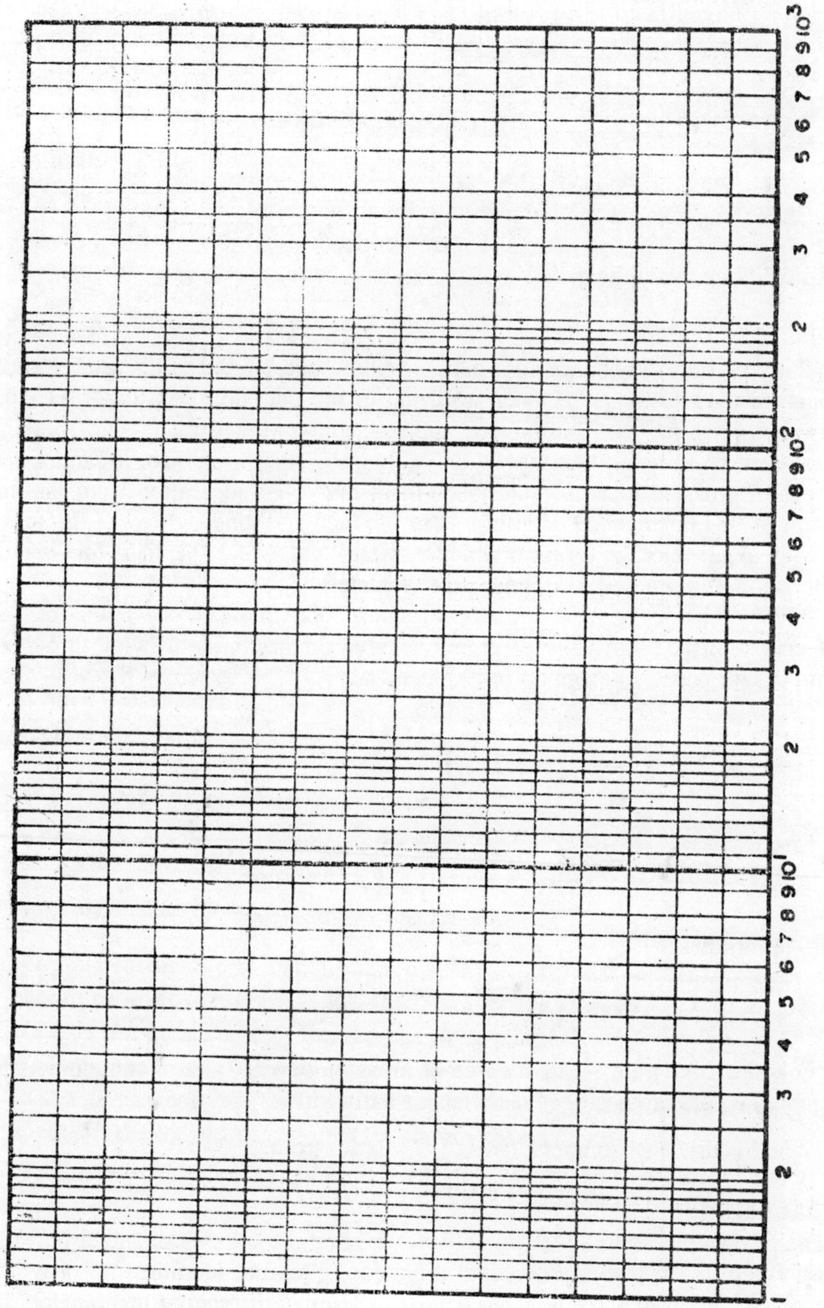

2.4 ELBOWMETER

Object

To calibrate the elbowmeter.

Theory

While the fluid flows around a bend or elbow, the pressure increases along the outer walls and decreases along the inner wall due to the curvature of flow. As the normal velocity and pressure distribution are approached on downstream side, the inner wall pressure must rise again. As the velocity along the surface is zero, it cannot decrease to provide for a pressure increase and hence there may be a separation from the inner wall. The result of this unbalanced condition is to produce a secondary flow. The motion combined with axial velocity forms a double spiral flow which persists for some distance downstream (approximately 100 times pipe diameter).

The dynamic or centrifugal force caused on the fluid between the inside and outside curves of the elbow, say $\triangle h$, is related to the discharge flowing through the pipe. Lansford has proposed an equation for 90° bend as

$$\left(\frac{p_0}{w} + Z_0\right) - \left(\frac{p_i}{w} + Z_i\right) = C_e \frac{U^2}{2g}$$

or $$\triangle h = C_e \frac{U^2}{2g}$$

Solving, for discharge

$$Q = C_d A \sqrt{2g.\triangle h}$$

where Q is the discharge, A is the area of cross section of inlet of the elbow and $\triangle h$ is the differential pressure head between outside and inside of the elbow.

The discharge coefficient C_d ($= \sqrt{1 C_e}$) varies from 0.55 to 0.9.

Experimental setup

The setup consists of a 90° pipe elbow (or bend) of suitable diameter (say, 50 mm or 75 mm), fitted in the pipeline of equal diameter. Pressure tappings are provided on the inner and outer walls of the eblow (or bend). A U-tube differential manometer is fitted to these points to measure the differential pressure. Water tank and a stop watch is used to measure the discharge (Fig. 2.4).

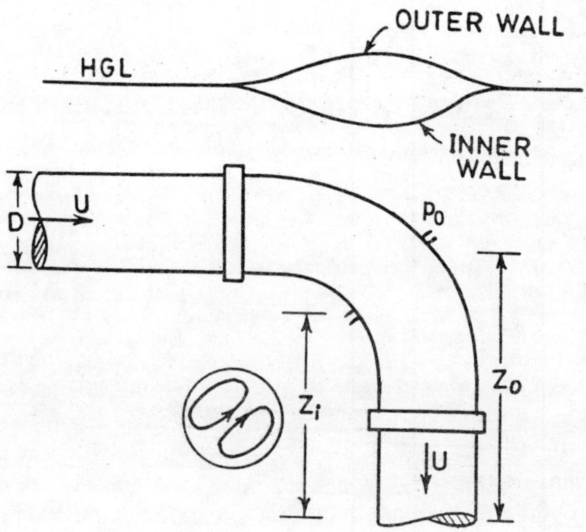

Fig. 2.4. Elbowmeter

Procedure

1. Regulate the inlet valve to allow a steady flow through elbowmeter.
2. Collect a certain volume of water (depth varying from H_1 to H_2) in the tank and note the time, t.
3. Note the manometer reading. Let the deflection between the levels of the manometeric liquid in the two limbs be, x.
4. Repeat steps 1 to 3 for other rates of flow.
5. Take five to six sets of readings

Calculations

1. The discharge Q is calculated by volumetric method. (Col. 5).
2. The deflection of mercury x (Col. 6) is converted into equivalent head of liquid of specific gravity S flowing in the pipe

$$\triangle h = x\left(\frac{S_m}{S} - 1\right) \qquad \text{(Col.7)}$$

3. The square root of $\triangle h$ is then computed in Col. 8.
4. The elbometer constant is computed as

$$C = \frac{\pi}{4} . d^2 . \sqrt{2g}$$

5. The coefficient of discharge is calculated for different sets.

$$C_d = Q/C\sqrt{\triangle h}$$

Observations

Diameter of pipe = m
Area of pipe A = m^2
Radius of pipe bend = m
Laboratory temperature = °C
Kinematic viscosity = m^2/s
Area of collecting tank A_1 = m^2

Sl.No.	Discharge reading Initial m	Final m	Time sec.	$Q = \frac{A_1 \Delta H}{t}$	Manometer deflection, x cm	Δh cm	$\sqrt{h}$	C_d	R_e	Log Re
1	2	3	4	5	6	7	8	9	10	11
							Average Value of C_d			

Presentation of results

1. Plot a graph between Q and $\sqrt{\triangle h}$ on an ordinary graph. The slop of this line is the coefficient of discharge multiplied by elbowmeter constant C.
2. Plot a graph between C_d versus log R_e.

Precautions

1. Remove all entrapped air from two limbs of manometer.
2. Maintain constant discharge for one set.
3. Take a number of readings to obtain accurate result.

Viva voce

1. What is the use of elbowmeter? Compare the merits of elbowmeter over an orficemeter.
2. Does the radius of curvature of the bend affect the coefficient of discharge?
3. Nearly 100 diameter lengths of the pipe should be placed on the downstream side. What is the idea?
4. Why is the pressure higher on the outside wall than on inside wall?
5. Draw HGL and TEL along the inner wall and outerwall of the elbowmeter.

2.5 ORIFICES

Object

1. To determine the coefficient of velocity and discharge for given circular orifice under constant discharging head.

2. To determine the coefficient of discharge for given non-circular orifice under varying discharge head (or time of emptying method).

Theory

1. An orifice is an opening in the side (or bottom) of a vessel and discharging the liquid contained in the vessel under pressure. The actual discharge Q is given by

$$Q = C_d\, a \sqrt{2g\,H}$$

With the help of equations of dynamics, the coefficient of velocity C_v is given by

$$C_v = \sqrt{x^2/4yH}$$

and coefficient of contraction C_c is given by

$$C_c = C_d/C_v$$

Here x and y represent the horizontal and vertical co-ordinates of jet and H is the constant head of water above the centre line of orifice.

2. If the head continuouslly falls from H_1 to H_2 in time interval t, then C_d is given by

$$C_d = \frac{2\,A}{a\,t\sqrt{2g}}\,(\sqrt{H_1} - \sqrt{H_2})$$

where A is the area of the tank and 'a' is the area of orifice.

Experimental set up

It consists of an inlet tank to which an opening is made on one of its walls near the bottom. In this opening, the given orifice can be fitted. (In a few set ups a revolving plate is fixed near the bottom. This revolving plate has openings of various shapes and sizes. This plate can be revolved so that desired orifice shape comes in front of the opening in the tank).

A horizontal scale with a hook gauge mounted on it is fixed to the tank just above the opening. Hook gauge can slide on the horizontal scale to measure the horizontal movement, (i.e. x-coordinate), the corresponding vertical coordinate (i.e. y) can be measured by hook gauge with reference to the section of vena contracta.

Procedure

CONSTANT HEAD METHOD

1. Fit the circular orifice in the opening of the tank and plug it.

2. **Fill the tank with water to a suitable** level and open the orifice.
3. **Adjust the inlet valve to make the water** level in the tank stand steady.
4. **Bring the hook gauge at the section of** vena contracta (near the orifice) and measure the coordinate x_1 and y_1 of a point on the jet.
5. Slide the hook gauge to any other point on the jet and measure corresponding points x_2, y_2.
6. Collect the discharge for certain time t in collecting tank of area A.
7. Repeat steps 2 to 6 for five different values of head H in the inflow tank.

VARIABLE HEAD METHOD

1. Replace the circular orifice with square/triangular orifice.
2. Plug the orifice and fill it to a desired level H_1.
3. Open the orifice and start the stop watch.
4. Take the water level reading H_2 and note the corresponding time t.
5. Repeat the steps 2 to 4 with five different initial and final heads over orifice.

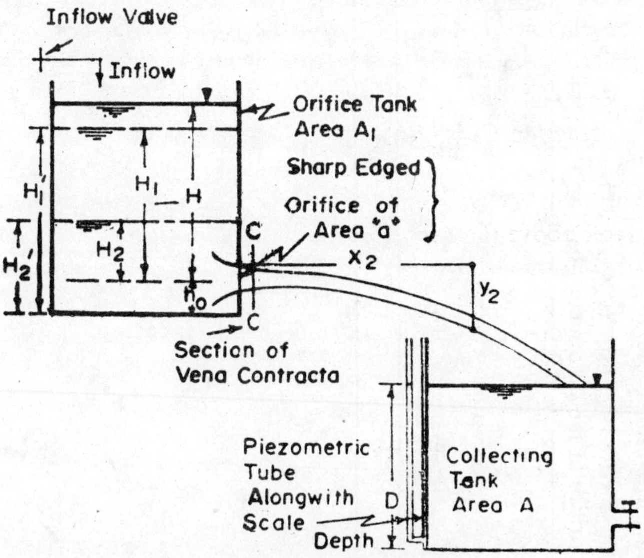

Fig. 2.5.1 Experimental set up

Calculations

CONSTANT DISCHARGE METHOD

1. Calculate the constant discharging head

$$H = H' - h_o \qquad \text{(col. 3)}$$

2. Calculate coefficient of velocity C_v

$$C_v = \sqrt{x^2/4\,y\,H} \qquad \text{(col. 6)}$$

3. Now calculate the coefficient of discharge

$$C_d = Q/a\sqrt{2g\,H} \qquad \text{(col. 11)}$$

Observations

(A) CONSTANT HEAD METHOD

Area of circular orifice 'a' = m^2

Area of collecting tank A = m^2

Initial reading on horizontal scale x_1 = m

Initial reading on vertical scale when hook gauge just touches the bottom of jet at vena contracta y_1 = m

Water level at the centre of orifice in the piezometer, h_o = m

Sl. No.	Head	H	x	y	C_v	Collecting tank reading			Q	C_d	C_c
	Final W.L. H' m	H m	$=x_2-x_1$ cm	$=y_2-y_1$ cm		Inital W.L. cm	Final W.L. cm	Time taken cm	m^3/sec		$=C_d/C_v$
1	2	3	4	5	6	7	8	9	10	11	12

(B) VARIABLE HEAD METHOD

Area of triangular/square/rectangular orifice = m^2

Area of orifice tank A = m^2

Piezometer reading of orifice tanks

water level at the centre of orifice h_o = m

Sl. No.	Orifice tank reading		Initial head	Final head	$\sqrt{H_1}$	$\sqrt{H_2}$	Time of emptying	Cd
	H_1'	H_2'	$H_1 = H_1' - h_0$ m	$H_2 = H_2' - h_0$ m			s	
1	2	3	4	5	6	7	8	9

4. Finally calculate the coefficient of contraction C_c

$$C_c = C_d/C_v \qquad \text{(col. 12)}$$

VARIABLE HEAD METHOD

5. The constant C is given as $2\,A/a\sqrt{2g}$
6. Calculate the initial head $H_1 = H_1' - h_o$ and final head as $H_2 = H_2' - h_o$ (col. 4 and 5)
7. Now compute the coefficient of discharge

$$C_d = C(\sqrt{H_1} - \sqrt{H_2})/t \qquad \text{(col. 9)}$$

Presentation of result

Find out the arithmetic mean values of C_v, C_d and C_c as their final values for circular orifice.

Coefficient of velocity, C_v =

Coefficient of contraction, C_c =

Coefficient of discharge C_d =

Similarly, the mean value of C_d for non circular orifice is determined.

Coefficient of discharge for triangular/rectangular orifice C_d =

Precautions

1. In case of circular orifice, take care that the head over the orifice is constant during all the time the readings are taken.
2. In case of noncircular orifice, take care that the noting of water level reading and stopping of stop watch take place at the same instant.

Viva-voce

1. What is an orifice ? If water stands only to the centre of it, will it still be called an orifice ?
2. Why is actual discharge less than theoretical discharge and how does C_d take this into account ? Can we obtain theoretical discharge in practice too ?
3. After changing the inlet valve opening, it is found that after some time, head again becomes constant at (of course) different level, why ?
4. What will be the effect on values of C_c, C_v and C_d if the edges of the orifice become eventually dull ?

2.6 MOUTHPIECES

Object

1. To determine the coefficient of discharge of the given cylindrical and convergent mouthpieces.

2. To show that the viscosity is a "superfluous variable" for flow through mouthpiece and thus to determine coefficient of discharge for given cylindrical and convergent mouthpieces.

Theory

The discharge through a mouthpiece is given by

$$Q = C_d\, a \sqrt{2g\, H}$$

where Q is discharge, a is exit area of the mouthpiece, H is head causing the flow (height of water level above the centre of mouthpiece) and C_d coefficient of discharge of mouthpiece.

If the phenomenon of flow through mouthpiece is to be analysed by dimensional analysis, the resulting equation will depend upon number of variables. Assuming that discharge Q through mouthpiece is function of density of water ρ (flowing through) diameter of the exit of the mouthpiece d, velocity through it U and also let us assume, viscosity of water μ, so that the starting equation is

$$Q = \phi\, (\rho, d, U, \mu)$$

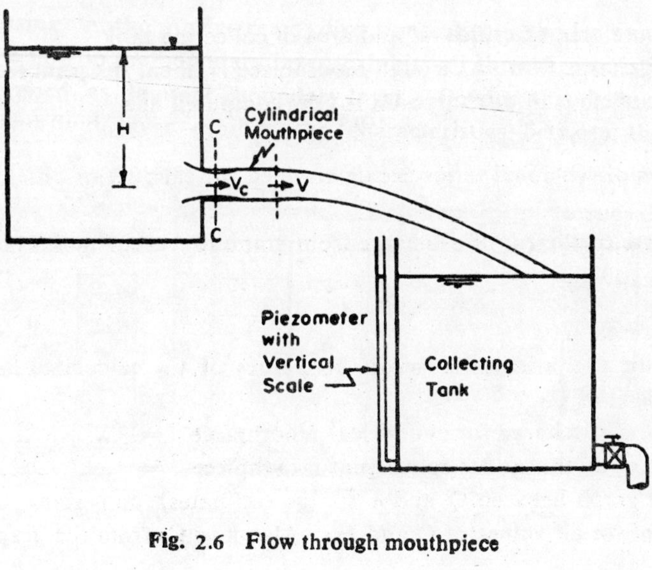

Fig. 2.6 Flow through mouthpiece

The above equation can be solved by dimensional analysis with the use of Bucking-ham's π theorem and incorporating that velocity U is not an independent variable but is proportional to $\sqrt{2g\,H}$, the following equation may be obtained.

$$Q/d^2\sqrt{gH} = \phi\,(\rho d\sqrt{gH}/\mu) = C$$

This equation can be plotted on a graph paper using experimental values. It can be shown that whatever the value of term containing viscosity, the value of other terms remains unchanged This means viscosity is a superfluous variable. Comparing

$$Q = C_d\, a\, \sqrt{2g\,H} = d^2\, \sqrt{gH}\, C$$

$$C_d = 0.9\ C.$$

Experimental set up

It consists of an inlet tank to which an opening is made on one of its sides (near the bottom). The mouthpiece is fixed in the opening. The collecting tank with a piezometer tube is fixed on one side of collecting tank along with a vertically mou ed scale (same used for orifices). A stop watch is also needed.

Proc re

1. Fit the required mouthpiece to the tank orifice and plug it.
2. Fill the tank to a particular level and then remove the plug.
3. Readjust the inlet valve to maintain constant discharging head. Measure the constant head.
4. Collect the flow in the collecting tank for a suitable time.
5. Repeat steps 2 to 4 with seven different heads.
6. Change the mouthpiece and repeat steps 2 to 5 for seven different heads.

Calculations

1. Calculate area of orifice 'a' and area of collecting tank.
2. The discharge flowing through mouthpiece is, then, determined. (col. 6)
3. The coefficient of discharge C_d is now computed as

$$Cd = Q/a\sqrt{2g\,H}$$

4. The non-dimensional terms are determined for each set of observation. (col. 8, 9)
5. Calculate coefficient of discharge from graph for each mouthpiece as given by $C_d = 0.9$ C.

Presentation of result

1. Calculate the arithmetic mean of values of C_d calculated in col. 7 as the final value of C_d.

Coefficient of discharge for cylindrical mouthpiece = ...

Coefficient of discharge for convergent mouthpiece = ...

2. Draw a graph between Q and $\sqrt{H}$ (as ordinates) and show that C_d is a constant for all values of Q and H. Also find C_d from the graph.

Observations

Exit diameter of mouthpiece 'd' = m

Area of collecting tank = m^2

Laboratory temperature = °C

Kinematic viscosity of water at this temperature = m^2/s

Sl. No.	Head H m	Collecting tank reading Initial W.L. m	Final W.L. m	Time sec m	Discharge Q m^3/s	C_d	$\frac{Q}{d^2\sqrt{gH}}$	$\frac{\rho d\sqrt{gH}}{\mu}$
1	2	3	4	5	6	7	8	9

3. Plot values of $Q/d^2\sqrt{gH}$ (col. 8) and $\rho\, d\, \sqrt{gH}/\mu$ (col. 9) separately for all mouthpieces on a graph. Show that former has a constant value C, whatever the values of the latter for each mouthpiece. This shows that viscosity is a superfluous variable. Also show that coefficient of discharge C_d for each mouthpiece is also given by

$$C_d = 0.9\, C$$

Precautions

Maintain constant water level in the orifice tank before each set.

Viva voce

1. Explain the hydraulic behaviour of different mouthpieces. How do they discharge more volume of water under same head and exit area as against an orifice. ?

2. What is a superfluous variable ? How is it detected with the help of hydraulic experimentation ?

2.7 MINOR LOSSES

Object

To determine the coefficients of minor losses for the pipe fittings.

Theory

The minor losses are given by the following expressions.

For right angled bend:

$$h_B = h_{L1,2} = C_B \frac{v^2}{2g}$$

For sudden enlargement of section:

$$h_E = h_{L2\,3} + \frac{v^2}{2g} - \frac{V^2}{2g} = C_E \frac{v^2}{2g}$$

For sudden contraction of section:

$$h_c = h_{3,4} + \frac{V^2}{2g} - \frac{v^2}{2g} = C_c \frac{v^2}{2g}$$

where v is velocity through the pipe of diameter 'd', V is velocity through the pipe of diameter 'D', C_B, C_E and C_c are coefficients of minor losses due to bend, sudden enlargement and sudden contraction respectively.

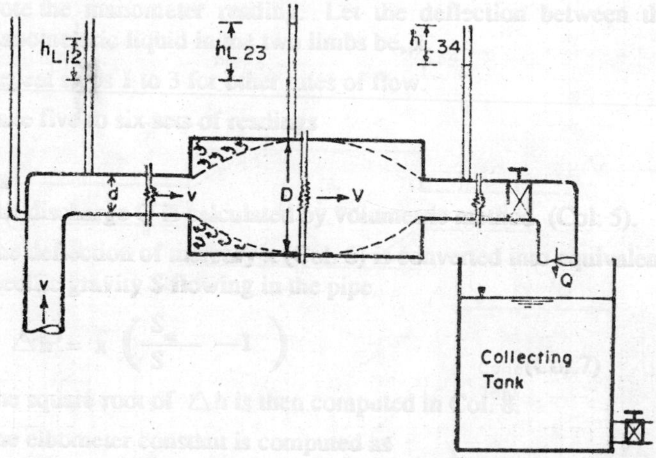

Fig, 2.7 Flow through a bend, sudden enlargement and sudden contraction

Experimental set up

It consists of a pipe 25 mm diameter which is fitted with a right angled bend, a sudden enlargement (another pipe of about 75 mm or so) and a sudden contraction.

Observations

Area of collecting tank $A =$ m^2

Area of small pipe $a_1 =$ m^2

Area of bigger pipe $a_2 =$ m^2

Sl. No.	Discharge readings Initial m	Final m	Time m	Discharge Q m^2/s	Manometer reading $h_{L1,2}$	$h_{L2,3}$	$h_{L3,4}$	Vel. v	Vel. V	$\frac{v^2}{2g}$	$\frac{V^2}{2g}$	C_B	C_E	C_C
1	2	3	4	5	6	7	8	9	10	11	12	13	14	15

This pipe discharges into a collecting tank used to collect a specific volume of water in time t. A stop watch is needed to record the time t.

Procedure

1. Allow a constant discharge to flow through pipe line.
2. Measure the discharge by volumetric method.
3. Take readings of U tube manometers connected across each of right angled bend, a sudden enlargement and sudden contraction.
4. Repeat the steps 1 to 3 for eight different rates of flow.

Calculations

1. Calculate the discharge Q (col. 5) and then determine the velocity u through small pipe of diameter d (col. 9) and also velocity U through large pipe of diameter D (col. 10).
2. Convert manometer readings into equivalent water heads (col. 6 to 8)
3. Calculate the velocity energy $u^2/2g$ and $U^2/2g$ (col. 11 and 12).
4. Now compute C_B, C_E and C_C by above equations.

Presentation of result

Average coefficient of various fittings are as follows:

1. Right angled bend C_B =
2. Sudden enlargement C_E =
3. Sudden contractions C_C =

Precautions

1. All the entrapped air should be driven out of legs of different manometers.
2. A constant discharge should be ascertained before taking the above readings.

Viva-voce

1. Why are losses due to bend, sudden enlargement and sudden contraction etc. known as minor losses ? What, then, are major losses ?
2. What is the advantage of finding out these coefficients ?
3. What causes a loss of head across a bend in a pipe line ?
4. In the given pipe line, if the direction of flow is reversed, what will be the effect on the values of differential mercury levels ?

2.8 MOMENTUM THEOREM

Object

To verify the momentum theorem, experimentally.

Theory

Whenever the velocity of a stream is changed either in magnitude or in direction, a force is required to bring this change.

Thus the momentum theorem states that the summation of all the externally applied forces on a given control volume is equal to the rate of change of momentum brought in the direction of forces.

$$\sum F_x = \rho Q\,(U_{2x} - U_{1x})$$

Where $\sum F_x$ is the summation of all the forces acting in x-direction. Similar equations can also be written in Y and Z direction.

By Newton's third law of motion, an equal and opposite force is applied by the fluid on the body which bring the change in momentum.

Since the jet of the fluid turns by right angle when striking a flat plate, the final velocity U_{2x} will be zero. Hence,

$$\sum F_x = \rho Q\, U_{1x}$$

Experimental setup

The setup mainly consists of a nozzle of about 10 mm in diameter fitted to a pipe of 50 mm diameter. The jet of water issuing from nozzle strikes a flat plate of about 100 mm diameter placed at some distance to the nozzle (Fig. 2.8). A force acts on the plate and by which the lever moves on the side. A weight W is placed on the pan P to counteract the momentum caused by the force F on the lever . The lever arm again comes to its original position. Thus, the force F acted by the jet on the flat plate equals the weight W.

The discharge, which falls in the tank after striking the flat plate is used to measure the discharge by volumetric method.

Procedure

1. Mark the position of vertical lever and horizontal arm when there is no weight on the pan.
2. Open the inlet valve so that jet of water strikes the flat plate at its centre. The lever is deflected to one side.
3. Put a weight W on the pan so as to bring the lever in its original position as marked in Step 1.
4. Measure the discharge Q by volumetric method.
5. Repeat steps 2 and 4 for various other discharges.

Observations

Diameter of nozzle = cm
Area of nozzle = m^2
Area of tank = m^2
Horizontal length of lever l_1 = m
Vertical distance of lever from the centre of flat plate t_2 = –m

Sl.No.	Discharge reading							
	Initial m	Final m	Time T	$Q = \frac{A_1 \triangle H}{t}$	Weight of the pan, W Newton	Equivalent force F	Force due to jet F_1	% error $\frac{F - F_1}{F} \times 100$
1	2	3	4	5	6	7	8	9

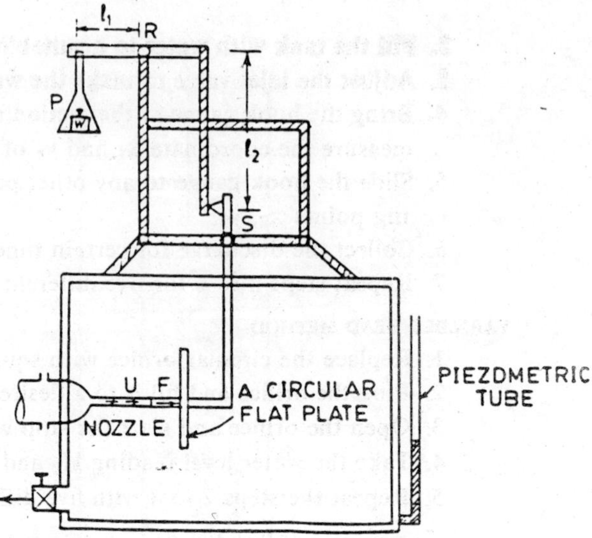

Fig. 2.8. Momentum theorem Set up.

Calculations

1. The discharge Q flowing out of nozzle is computed in Col.4.
2. The moment of W and F about the lever arm hinge R yields

$$W. l_1 = F_1 \ l_2$$

or $$F_1 = W \ l_1/l_2 \qquad \text{(Col.8)}$$

3. Theoretical force imparted by the jet on the flat plate is given by

$$F = \frac{w}{g} Q. U = \frac{w}{g} \cdot \frac{Q^2}{a} \qquad \text{(Col.7)}$$

4. Percentage error is determined as $F — F_1/F \times 100$ in Column 9.

Presentation of result

1. Plot F (as abscissa) versus F_1 (as ordinate). It should be a straight line inclined at 45° to the x-axis. Thus it proves the theorem.

Precautions

1. The discharge should remain constant in one set of reading.
2. The jet issued from the nozzle should be solid and no water should spray in the atmosphere, so that the whole mass (fluid) impinges on the flat plate.
3. The hinges R and S should work smoothly.

Viva-voce

1. What is momenturm theorem? State the same.
2. Compare the energy principle and momentum principle.
3. Mention a few problems which can be solved by the application of momentum theorem.

2.9 PIPE FRICTION

Object

1. To determine Darcy's friction coefficient 'f' for (two) given pipes.

2. To obtain log f vs. log Re graph for (two) different pipes and mention the value of κ/D on each.

Theory

As the fluid (here water) flows through a pipe, the viscosity of fluid and surface (inside) or pipe offer resistance to the flow, and in overcoming the same, energy of the flowing water is lost. This loss of energy per unit weight of water, called loss of head due to friction 'h_f', is given by the following Darcy's formula

$$h_f = f\frac{L}{D} \times \frac{U^2}{2g} \qquad (1)$$

where U is the mean velocity of fluid in the pipe of diameter D and length L (across which h_f was measured) and f is known as 'Darcy's friction coefficient'. For a horizontal pipe with constant discharge, this loss of head will appear as the difference of pressure at two sections L apart.

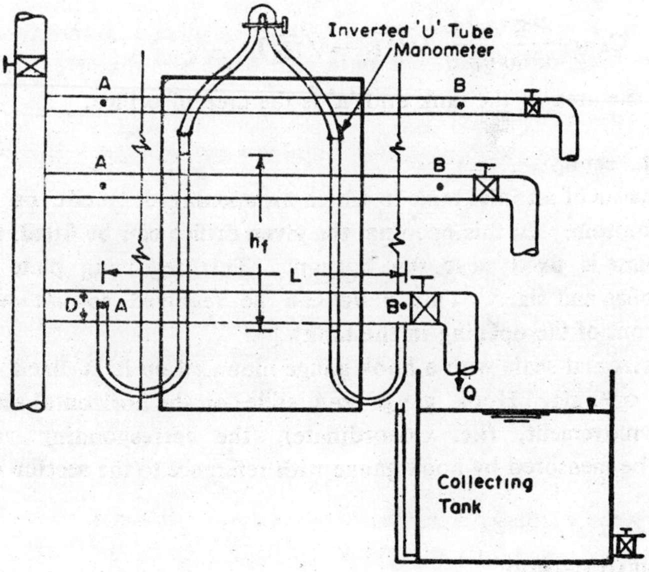

Fig. 2.9.1. Experimental set up

When a real fluid flows through a pipe, it has been experimentally shown by Reynolds, there exist two distictly different regimes of flow under. Of course, under a set of circumstance—these are, Laminar flow, usually at low velocities and Turbulent flow, at higher velocities, Laminar flow has little significance. Turbulent flow has different stages of flow, three sub regions of flow, namely smooth turbulent flow rough turbulent flow and transition between smooth and turbulent flow. A complete dimensional analysis of pipe friction phenomenon on comparison with Darcy's equation (Eq. 1 above), reveals that Darcy's friction factor f depends on (i) Renolds numer and (ii) relative roughness, κ/D where κ is the average height of surface roughness of pipe. κ is defined as the diameter of such uniform sand grains, which when coated on a pipe wall, yield the same limiting value, for rough conditions, as given by the pipe.

$$\text{Reynolds number, } R_e = \rho UD/v = 4Q/\pi Dv \tag{2}$$

A graphical plot between log f and R_e for different values of κ/D will offer a complete revelation of the nature of pipe friction phenomenon for all the three sub regions of turbulent flow, Fig. 2.8.2. It will be found that the value of 'f' depends on (1) R_e alone in smooth turbulent flow, (2) κ/D alone in rough turbulent flow and (3) κ/D and R_e both in transition flow. The value of f for rough turbulent flow is given by equation

$$1/\sqrt{f} = 1.14 - \log \kappa/D \tag{3}$$

This equation can be used to find κ provided 'f' is known and flow is definitely rough turbulent flow.

Experimental set up

The set up, Fig. 2.8.1 consists of a number of pipe, 12.5 mm to 38 mm diameter and nearly 5 m long, connected to a main pipe of 38 mm diameter. An inlet valve is

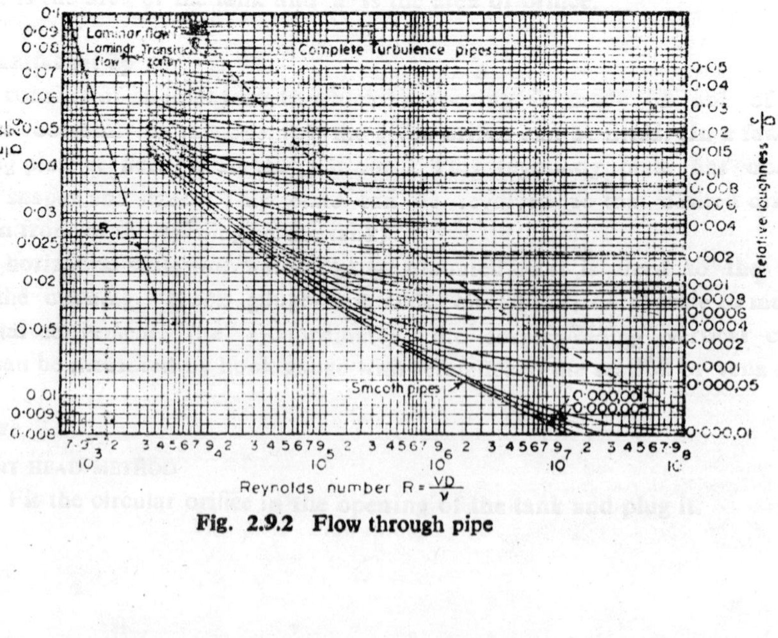

Fig. 2.9.2 Flow through pipe

provided in the main pipe to regulate the discharge to different pipes. A valve is also fitted at the outlet of each pipe to regulate the flow in it. A U-tube manometer may be fitted between two suitably located pressure tappings A and B on each pipe to measure the head loss h_f. A collecting tank is required to measure the discharge.

Procedure

1. Check the pressure connections to the pipe in question and determine its length and diameter (col. 1)
2. Allow a small discharge to flow through the pipe and note the pressure difference (col. 7) when it becomes constant.
3. Collect the flow in a collecting tank for a suitable time and note the initial and final water levels and the time taken (col. 3, 4, 5).
4. Note the manometer reading (col. 7).
5. Change the delivery valve opening and adjust another discharges.
6. Repeat the above procedure for different discharges.
7. Follow similar procedure for successive pipes.
8. Record the laboratory temperature and hence corresponding value of kinematic viscosity.

Calculations

1. Calculate the discharge Q and hence corresponding mean velocity U (col. 6, 8).
2. Now calculate friction coefficient f using Eq. 1.
3. Calculate Reynolds number Re (col. 12).
4. Determine log f (col. 11) and log Re (col. 13)

Presentation of result

1. Find for each pipe the arithmetic mean of the values of f (col. 10) as its final value. Draw a graph between h_f (col. 7) and U^2 (col. 9) and show that f is a constant for all velocities. Also find from the graph the value of 'f'

f for..............m diameter pipe =

f for..............m diameter pipe =

2. Plot the values of log f (col. 11) as ordinate against log Re as abscissa separately for different pipes (on the same graph). Identify the portions of curves representing different types of flow. For two values of f corresponding to maximum discharges through each pipe and, using Eq. 3, calculate the value of κ/D. Mention the same on the corresponding curve in the graph in the rough turbulent flow zone.

κ/D for..............in diameter pipe =

κ/D for..............in diameter pipe =

Observations

Area of collecting tank A = m^2

Length of pipe L = m^2

Laboratory temperature $T°$ = °C

Kinematic viscosity ν at $T°C$ = m^2/s

Sl. No.	Dia of pipe m	Discharge reading Initial m	Final m	Time sec	Q m^3/s	h_f m	U m/s	U^2	f	log f	Re	log Re
1	2	3	4	5	6	7	8	9	10	11	12	13

Viva-voce

1. Why does the pressure along a horizontal pipe go on decreasing?

2. Is the Darcy's friction coefficient really a constant? On what factors does it depend?

3. What are hydraulic gradient line and total energy line? A constant discharge flows through a long gradually converging pipe line. Do you expect a falling or rising hydraulic gradient line?

4. Discuss a double log graph between f and R_e. Identify therein the three different sub regions of turbulent flow and friction loss for each sub region.

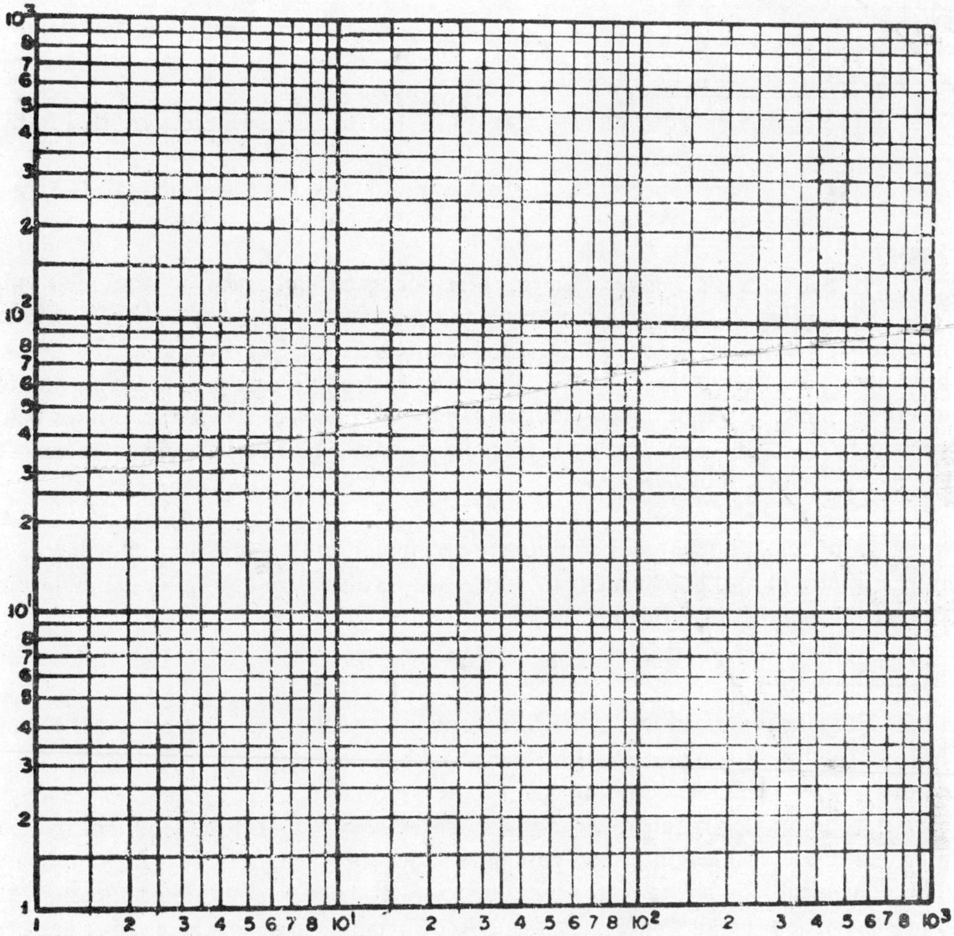

2.10 BYE-PASS

Object

1. To determine the discharge coefficient of bye-pass, (Q/Q_b).
2. To determine Darcy's friction coefficient for bye-pass.
3. To determine 'equivalent length' of bye-pass for the watermeter fitted in bye-pass.
4. To determine constant 'C' if the loss in the meter is assumed as $C\ U_1^2/2g$.

Theory

The bye-pass is a smaller diameter pipe connected in parallel to the main pipe for a short length. The two junctions are connected to the two ends of a pressure manometer. Pressure head reading of the manometer, say H, is equal to the loss of head due to friction in the main pipe as well as in the bye-pass pipe between the two junctions, minor losses at the junctions having been neglected. All quantities below pertaining to main and bye-pass pipes have been written with suffixes 'm' and 'b'.
Discharge coefficient of bye-pass

$$K_D = Q/Q_b \tag{1}$$

where Q is total discharge pertaining to the junction, Q_b is discharge flowing in the byepass pipe and Q_m which is equal to $Q - Q_b$ is discharge flowing in the main pipe. Then, from parallel pipe theory,

$$H = f\frac{l_m}{d_m}\,\frac{U_m^2}{2g} = f\frac{l_b}{d_b}\,\frac{U_b^2}{2g} = f\frac{l_b}{d_b}\,\frac{U_b^2}{2g} + C\,\frac{U_b^2}{2g} \tag{2}$$

where H is pressure head lost in friction in main as well as bye-pass pipe, f is Darcy's friction coefficient assumed same for main and bye-pass pipes, l_m, d_m, U_m are length diameter and mean velocity of main pipe respectively, l_b, d_b, and U_b are length, diameter and velocity of bye-pass pipe respectively. L_b is total length of bye-pass pipe and is equal to length of bye-pass pipe+equivalent length (say l_e) for watermeter. This equivalent length means that loss due to watermeter is equivalent to loss of head due to friction in the equivalent length of bye-pass pipe in which meter has been fitted. From first part of Eq. 2, H, l_m, d_m, U_m being known, 'f' and L_b of bye-pass can be calculated. But $L_b = l_b + l_e$, thus l_e can be calculated. From third part of Eq. 2, it is being evident that constant $C = (f\ l_e/d_b)$ can be calculated.

Experimental set up

It consists of a bigger diameter pipe (say; 25 mm diameter pipe), having inlet valve at upstream end. A smaller pipe (of 12.5 mm diameter) is connected in parallel to main pipe. The head loss is measured by means of an inverted U—tube manometer and the discharge flowing out is measured by volumetric method.

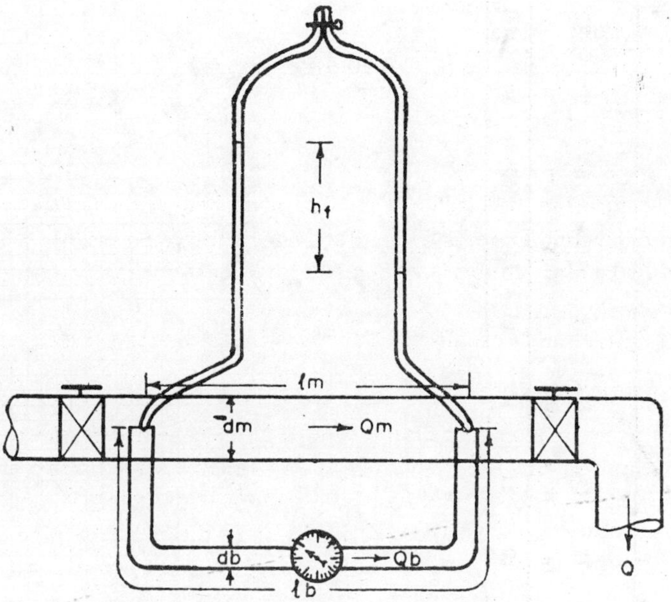

Fig. 2.10 Flow through bye-pass

Procedure

1. Open the valve and maintain a constant discharge through the main pipe.
2. Collect volume of water in the tank (col. 2 and 3) for a suitable time (col. 4), compute discharge 'Q' (col. 5).
3. Note the initial and final readings of watermeter after a suitable interval of time (col. 6 and 7).
4. Observe the reading of manometer (col. 10)
5. Repeat the above steps for eight sets of dischages.

Calculations

1. Discharge flowing through main pipe $Q_m = Q - Q_b$
2. The friction coefficient for bye-pass $f_b = f_m$

   $f_m = Hd_m\ 2g/l_m\ U_m^2$
3. The total length of bye-pass $L_b = Hd_b\ 2g/f\ U_b^2$
4. Equivalent length of bye-pass pipe is then found out by substracting the actual length of bye-pass l_b from L_b

   $l_e = L_b - l_b$
5. Finally the constant is then computed as

$$C = \frac{2g}{U_b^2}\left(H - \frac{f\ l_b}{d_b}\ \frac{U_b^2}{2g}\right) = \frac{f\ l_e}{d_b}$$

Observations

Length of main pipe L_m = m
Diameter of main pipe d_m = m
Length of bye-pass pipe l_b = m
Diameter of bye-pass pipe d_b = m
Area of collecting tank A = m^2

Sl. No.	Discharge Reading Initial m	Final m	Time sec	Q m^2/s	Bye-pass discharge Watermeter reading	Time sec	Q m^3/s	K_D	H m	Q_m m^3/s	U_m m/s	f	U_b m/s	l_b m	l_e m	C
1	2	3	4	5	6	7	8	9	10	11	12	13	14	15	16	17

Presentation of results

1. Find the arithmetic mean of the values of K_D (col. 9) as its final value. Draw a graph between Q and Q_b to show that K_D is constant.
   The coefficient of discharge for bye-pass K_D = ...
2. Find the arithmetic mean of the values of Darcy's f (col. 13) for bye-pass as its final value.
   Friction coefficient f =
3. Find the arithmetic mean of the values of equivalent length l_e (col. 16) as its final value for the watermeter.
   Equivalent length = ...m
4. Find the arithmetic mean of the values of 'C' (col. 17).
   The value of loss coefficient 'C' = ...

Precaution

1. Observe the values of Q_b carefully, otherwise it will affect the value of K_D considerably.
2. Inlet valve should not be touched during one set of reading.

Viva-voce

1. What is a bye-pass ? What is the use of this experiment in practice.
2. How will you determine the discharge flowing past a 2 m diameter water supply main ?
3. If the direction of flow is reversed and same difference of pressures is maintained between A and B, will the valve of Q_b change ? How ?

2.11 CRITICAL REYNOLDS NUMBER

Object

1. To study the transition of laminar to turbulent flow.
2. To obtain the lower critical Reynolds number.

Introduction and theory

A real fluid is characterized by the presence of viscosity. Depending on the relative magnitude of viscous and inertia forces, the flow can take place in two entirely different regimes—laminar and turbulent flows.

At low velocities, the viscous forces are predominant than inertia forces, the fluid appears to move in layers (or laminar). Such a flow is designated as laminar flow. On the other hand, as the velocity of flow increases, the inertia forces overcome the viscous forces and become predominant. The fluid layers get mixed with each other in a rather chaotic manner. Such a flow is termed as turbulent flow. This fact can be determined in a Reynolds apparatus.

The distinction between these two regimes of flow can also be demonstrated by plotting the relationship between energy lost during flow, h_f, versus mean velocity (Fig. 2.11).

The relationship between h_f and U is found to be

$$h_f \propto U^n$$

In the earlier portion of the plot when the velocities are small, it is noticed that h_f is directly proportional to velocity U, (i.e. for laminar flow, n = 1). In the latter portion of the

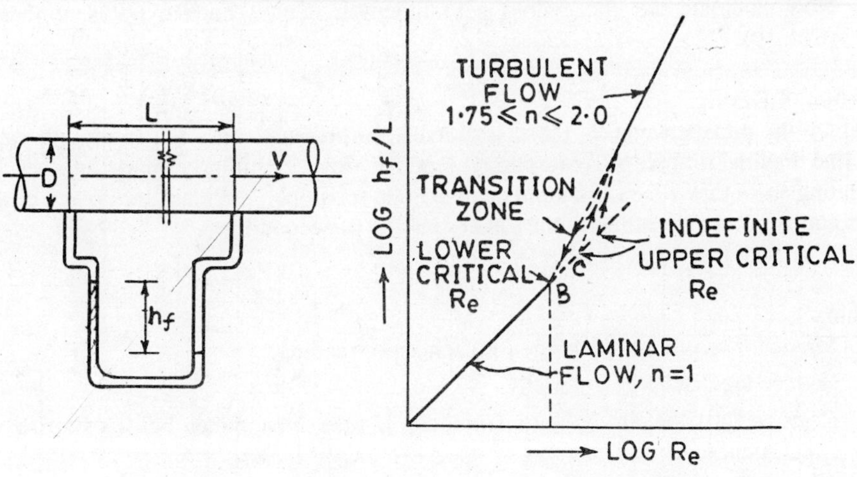

Fig. 2.11. Experimental setup and dimensionless plot.

curve, it is noticed that the value of exponent n lies between 1.75 to 2. (Generally n is assumed to be equal to 2 for turbulent flow). Usually a dimensionless plot between h_f/L versus Reynolds number, UD/ν is plotted on a log-log paper. Here h_f is the energy loss occurring due to flow of water at mean velocity U in a pipe of length L and of diameter D.

It is to be observed from this plot that if the flow is increased from its lower value to higher, the transition takes place at point C. But if the flow is decreased from its maximum value, it does not follow the same trend but the transition of flow takes place at another point B. The Reynolds, number corresponding to the point where flow characteristics start changing is termed 'Critical Reynold's Number' (R_e) critical. The value of R_e corresponding to C is known as 'Upper Critical Reynold's Number' and R_e corresponding to B is known as 'Lower Critical Reynold's Number'. Generally, the lower critical Reynold's number is taken as 2000 and is of much practical importance.

Experimental setup

It consists of a pipe (10 mm to 15 mm diameter) of sufficient length (say, 10 m to 12 m) to which two pressure tappings are fitted at suitable spacings. An inverted U-tube manometer is fitted to these pressure tappings. The discharge is regulated by a sluice valve and is measured by volumetric method.

Procedure

1. Set the U-tube manometer properly and allow the maximum discharge to flow.
2. Note the pressure difference between two pressure points.
3. Collect the water in a collecting tank for a suitable interval of time.
4. Repeat steps 2 and 3 with gradually decreasing discharge (in steps).

Computation

1. Calculate the discharge Q and head loss h_f between two pressure points (Col. 5 and 8).
2. Now calculate the energy loss per unit length. (Col. 9) and Reynolds number, R_e (Col. 10).

Presentation of Result

Prepare the plot between R_e (abscissa) and h_f/L (ordinate) on a log-log graph paper. Mark a line having unit slope representing laminar flow. Similarly mark another line corresponding to higher Reynolds number and finds its slope. Indicate the lower critical Reynolds number (corresponding to the lower extreme of second line).

Lower critical Reynolds number =

Precautions

1. Maintain a constant discharge for each set of reading.
2. Reduce the discharge gradually.
3. Remove all the air bubbles from the U-tube manometer before starting the experiment.

Observations

Diameter of pipe	=	m
Length of pipe between pressure tappings	=	m
Temperature of laboratory	=	°C
Kinematic viscosity of water	=	m^2/s
Area of collecting tank	=	m^2

Sl.No.	Discharge reading			Q	Manometer reading		h_f	$\frac{h_f}{L}$	$R_e = \frac{4\rho Q}{\pi \mu D}$
	Initial m	Final m	Time T	m^3/s	Left side	Right side			
1	2	3	4	5	6	7	8	9	10

Viva-voce

1. What is the practical utility of lower and upper critical Reynolds number?
2. What is Reynolds number? How is it useful in distinguishing the regime of flow?
3. What will happen if the water flows with initial disturbance?
4. How will the shape of entry to the pipe affect the lower critical Reynolds number? Will it be more or less?

2.12 STOKES LAW

Object

1. To verify the Stokes law.
2. To study the variation of coefficient of drag with Reynolds number.

Introduction and theory

When a solid body moves through a fluid of infinite extent, it experiences a resistance which acts in the direction opposite to the motion of body. This resistance depends upon the size of the body, the velocity with which it moves, the viscosity of the fluid and its mass density. Sir G. Stokes (1819-1903) studied the uniform motion of a sphare through a large extent of viscous fluid at very small Reynolds number. He gave the expression for drag F_D.

$$F_D = 3\pi \; D \mu U \tag{1}$$

where D is the diameter of the sphere and U is the mean velocity of the body through the stationary fluid. The only force overcoming the viscous force is the effective weight of the sphere, i.e.

$$F_D = \text{Weight of sphere} — \text{Buoyant force due to fluid.}$$

$$= \frac{\pi D^3}{6} (w_s — w) \tag{2}$$

where w_s and w are specific weight of solid and fluid respectively.

Equating Eqs. (1) and (2)

$$U = \frac{D^3}{18 \mu} (w_s — w) \tag{3}$$

The drag force F_D exerted by the sphere can also be expressed as

$$F_D = C_D . A . \frac{\rho U^2}{2} \tag{4}$$

In the expression C_D is known as coefficient of drag. A is the projected area of the body at right angle to flow and ρ is the mass density of fluid.

Equating the two valves of F_D in Eqs. (1) and (4),

$$C_D = \frac{24}{R_e} \tag{5}$$

Equation (5) is valid for Reynolds number upto 0.1.

If the spherical particle of diameter D is moving in finite fluid medium, the velocity U of the sphere (Eq. 3) should be corrected for wall constraint. The corrected velocity U_1 is given by

$$U_1 = U \left(1 + 2.4 \frac{D}{D_1}\right) \qquad (6)$$

where D_1 is the diameter of the container used in the experiment.

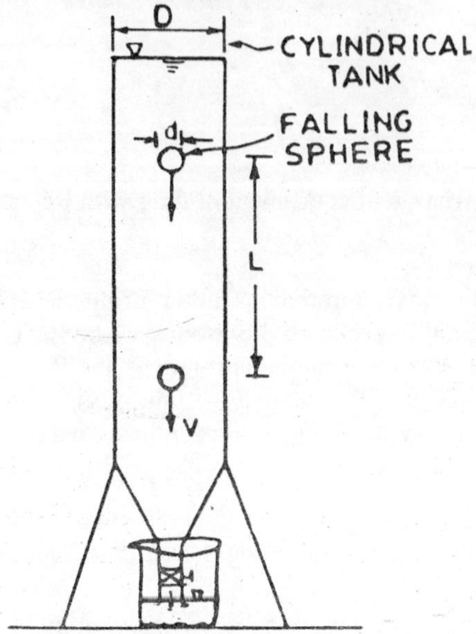

Fig. 2.12.1. Experimental setup.

Experimental setup

It consists of a cylindrical container of about 200 mm diameter and 2m height. A hopper and a valve is provided at its bottom to collect the steel balls. A transparent strip

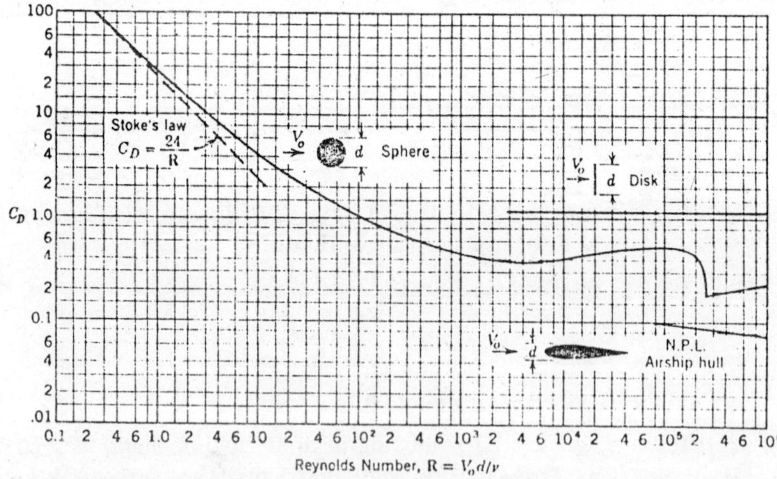

Fig. 12.2.2. Variation of coefficient of drag with Reynolds number for 3 D bodies.

of perspex is fitted throughout its height to observe the motion of the sphere. The container is filled with viscous fluid, e.g. glycerine, the viscosity of this liquid is determined using Redwood viscometer, or Saybolt viscometer. A number of small diameter balls, a stop watch and a metric scale are needed for conducting the experiment. A micro vernier is used to measure exactly the diameter of spherical balls.

Procedure

1. Measure the diameter of balls at diametrically opposite points.
2. Determine the viscosity of the fluid at laboratory temperature using Saybolt/ Redwood viscometer or from standard tables.
3. Fill the container with viscous fluid upto 10 cm from the brink.
4. Hold the ball by hand near the liquid surface and leave them gently.
5. Note down the time taken by the balls to travel 1 m depth.
   Note : Start the stop watch after the ball has travelled 150 mm or so.
6. Repeat steps 4 and 5 with other balls.
7. Collect the balls from the bottom of container by opening the valve.

Determination of viscosity

8. Take 100 c.c. of liquid and pour it in the internal container of viscometer and adjust the initial level of liquid in the reservoir to standard height.
9. Open the valve at the bottom of reservoir and start the stop watch.
10. Collect 60 c.c. of liquid in a container and note drown the time required for it. Now calculate the viscosity of liquid using Eq. 7.

$$\gamma = 0.0022t - \frac{1.8}{t} \tag{7}$$

Computations

1. Calculate the drag force F_D, using Eq. 2.

$$F_D = \frac{\pi D^3}{6} (w_s - w)$$

2. Find out the terminal velocity of balls,

$$U = L/T \tag{8}$$

3. Determine the corrected terminal velocity of balls

$$U_1 = U\ (1 + 2.4\ D/D_1) \tag{9}$$

4. Now compute the coefficient of drag C_D,

$$C_D = \frac{4F_D}{\pi D^2 \quad \rho U_1^2/2} \tag{10}$$

5. Determine the Reynolds number corresponding to the corrected terminal velocity of spherical balls.

Presentation of results

1. Plot the values of C_D (ordinate) and R_e (abscissa) on a log-log paper. In the range of $R_e \leqslant 0.1$, the experimental points should fit the line given by Eq. (5) and state the same.

   The points fit the line ($C_D = 24/R_e$) = Yes/No.

Precautions

1. The balls should be placed on the liquid surface gently.
2. Let the balls travel a small distance (say 150 mm) and acquire terminal velocity.

Viva-voce

1. What will be the difference in the coefficient of drag if a flat circular disc is used instead of a spherical balls?
2. What will be the effect of temperature on the coefficient of drag?

2.13 BOUNDARY LAYER

Object

1. To determine the boundary layer thickness, displacement boundary layer thickness and momentum boundary layer thickness.
2. To determine the boundary shear stress.
3. To determine the exponent in the power law of velocity distribution.

Theory

When the real fluid flows past a body, the fluid particles stick to the surfacc due to its viscosity and acquires the velocity of contact surface i.e. no slip condition prevails between the contact surface and the fluid layer nearest to it. The fluid velocity increases gradually and becomes nearly equal to free stream velocity away from the boundary. The region of flow in which this change of velocity takes place is known as boundary layer. Away from boundary, the ambient velocity U_o obtained is termed free stream velocity.

Consider a flat plate held parallel to the flow in a uniform stream of velocity U_o. As the fluid reaches section 1, the velocity of fluid at the boundary becomes zero and increases asymptotically to ambient velocity U_o away from the place and hence a velocity gradient is set up which develops shear resistance. Near the leading edge of the plate, it is only in a very thin layer that the fluid is retarded. As this retarded fluid layer moves downstream, continued action of shear resistance causes retardation in the increasing quantities of fluid. As a result, the thickness of boundary layer goes on increasing in the downstream direction. From the leading edge upto a certain length, the flow in the boundary layer exhibits all the characteristics of laminar flow and is termed as laminar boundary layer. Beyond this distance, the flow becomes unstable and fluid layer exhibits the characteristics between those of laminar and

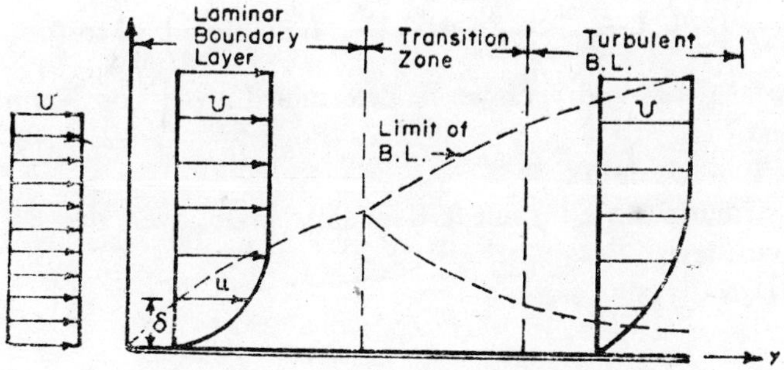

Fig. 2.10.1 Boundary layer over flat plate

turbulent motions and hence is known as transition region. At the downstream of this transition region, the flow in the boundary layer becomes turbulent and this portion of boundargh layer is known as turbulent boundary layer. For practical purposes, the critical Reynolds number (at which transition from laminar to turbulent boundary layer occurs) can be taken as 5×10^5.

Nominal thickness of boundary layer: It is the distance normal to the surface in which the fluid acquires the velocity 1% less than the ambient velocity (free stream velocity) i.e. it is equal to δ = (at which $u = 0.99\ U_o$)

Displacement thickness of boundary layer: It is defined as the distance the actual boundary would have to be displaced in order that actual discharge would be the same as that of an ideal fluid past the displaced boundary.

$$\delta_* = \int_o^\infty \left(1 - \frac{u}{U_o}\right) dy = \int_o^\infty \left(1 - \frac{u}{U_o}\right) dy \qquad (1)$$

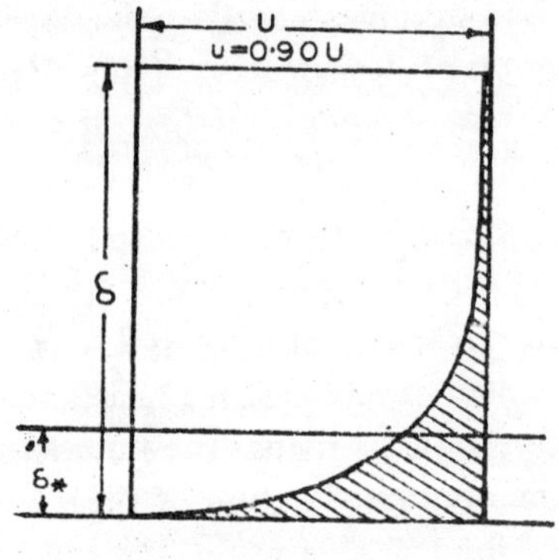

Fig. 2.10.2 Boundary layer thickness

Momentum thickness of boundary layer: It is defined as the distance from the actual boundary such that the momentum flux through this distance is the same as the deficit of momentum flux due to boundary layer formation.

$$\phi = \int_o^\infty \left(1 - \frac{u}{U_o}\right) \frac{u}{U_o} dy = \int_o^\delta \left(1 - \frac{u}{U_o}\right) \frac{u}{U_o} dy \qquad (2)$$

The shear stress at the boundry τ_o can be determined using Von Karmon momentum integral equation.

$$\tau_o/\rho \simeq U_o\, d\theta/dx \qquad (3)$$

The velocity distribution in a turbulent boundary layer over the flat plate can be expressd by power law

$$u/U_o = (y/\delta)^n \qquad (4)$$

Procedure

1. Start the wind tunnel and ensure that the pressure gradient is zero in the wind tunnel.

Observations

Specific weight of liquid	w =	km/m^3
Slope of inclined manometer	a =	
Specific weight of air	w_a =	km/m^3
Conversion factor for converting manometer reading	C =	
Diameter of Prandtl's tube	d =	mm

Sl. No.	Prandtl's tube Reading: Initial	Prandtl's tube Reading: Final	Prandtl's tube Reading: Dist. from bottom y mm	Manometer Reading: Left limb	Manometer Reading: Right limb	Manometer Reading: Differ. x	u	$\frac{u}{U_o}$	$\frac{y}{\delta}$
1	2	3	4	5	6	7	8	9	10

2. Place the Prandtl's tube at the tunnel bottom and take the manometer reading.
3. Raise the Prandtl's tube by 2 mm and repeat step 2.
4. Repeat the steps 3 till centre of wind tunnel is reached or no change in the manometer reading is noted.

Note : Use inclined U-tube manometer or electronic instrument, (pressure transducers) for pressure measurements.

Calculations

1. Calculate the velocity 'u' of fluid at any distance, y, from tunnel boundary.
   $u = C\sqrt{2g.\ \Delta x}$ (col. 8)
2. Plot the velocity u versus y and determine the free stream velocity U_o.
3. Determine the distance δ where $u=0.99\ U_o$ which is nominal B.L. thickness.
4. Find the area of the shaded portion of the velocity defect profile (Fig. 2).
   $\delta_* = \text{Shaded area}/U_o$
5. Plot u/U_o vs. y. Divide this plot into small strips and determine the product $(u/U_o)\ (1-u/U_o)\ dy$
6. Add all such products
   $\theta = \int u/U_o\ (1-u/U_o)\ dy =$
7. Plot u/U_o vs y/δ on a log-log graph. Fit in a straight line and determine its slope.
8. Plot θ verses X.
9. The wall shear is determined from Eq. 3.

Presentation of results

Nominal boundary layer thickness $\delta =$
Displacement boundary layer thickness $\delta_* =$
Momentum boundary layer thickness $\theta =$
Shear stress $\tau_o =$
Exponent 'n' = slope of log u/U_o vs. log y/δ plot =

Precautions

1. Read the inclined manometer correctly.
2. Draw the smooth curve between θ and X.

Viva-voce

1. Define boundary layer thickness—δ, δ_* and θ.
2. What will be the effect of pressure gradient on the growth of boundary layer ?

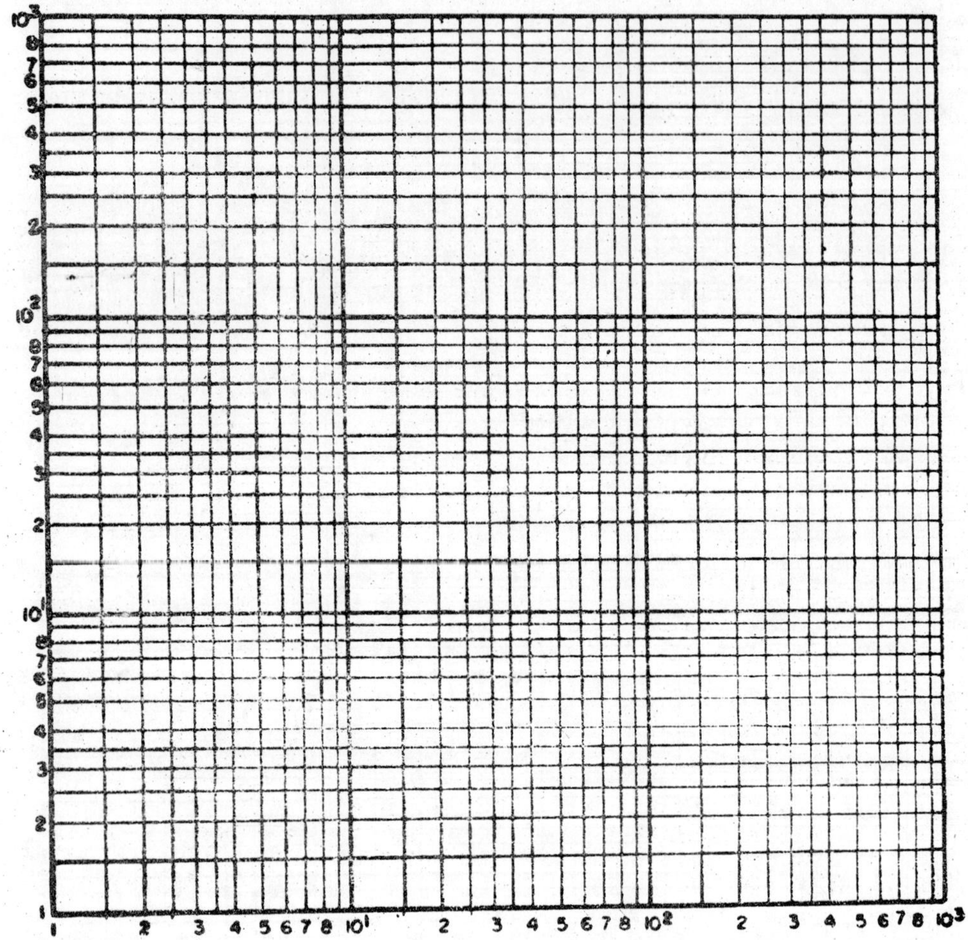
1 2 3 4 5 6 7 8 10¹ 2 3 4 5 6 7 8 10² 2 3 4 5 6 7 8 10³

2.14 FLOW AROUND CIRCULAR CYLINDER

Object

To measure the pressure distribution around a cylinder placed in the air stream and calculate coefficient of drag therefrom.

Theory

When an object is placed in the flowing fluid, it experiences a resistance F. The component of this force in the direction of flow is known as the drag force F_D The drag force consists of deformation drag, form drag and friction drag.

1 FRICTION DRAG The resistance experienced by the body due to the fluid viscosity and surface roughness is known as friction drag.

2 DEFORMATION DRAG It is caused due to the deformation of flow field in the body. Deformation drag consists of friction drag at the boundary and pressure drag due to the variation of pressure caused by the deformation of the flow pattern. The pressure drag reduces with increase in Reynolds number.

3 FORM DRAG At high Reynolds number, if the shape of the body is such that the separation of flow occurs, the pressures will be lower on the rear side than on front portion. This results in an additional force acting on the body in the direction of flow and is termed as form drag.

Consider a body placed in the flow, mean velocity being U. The total drag force acting on it is F_D which can be resolved into two components-one tangential to the body or friction drag F_f and other normal to it or pressure drag F_p The pressure and viscous stresses acting on an elementary area dA are p and τ_0 respectively.

$$\therefore dF_D = p \cos\theta \, dA + \tau_0 \sin\theta \, dA$$

Hence total drag acting on the whole body is the summation of dF_D

$$\therefore F_D = \int_A p \cos\theta \times dA + \int_A \tau_0 \sin\theta \times dA \qquad (1)$$

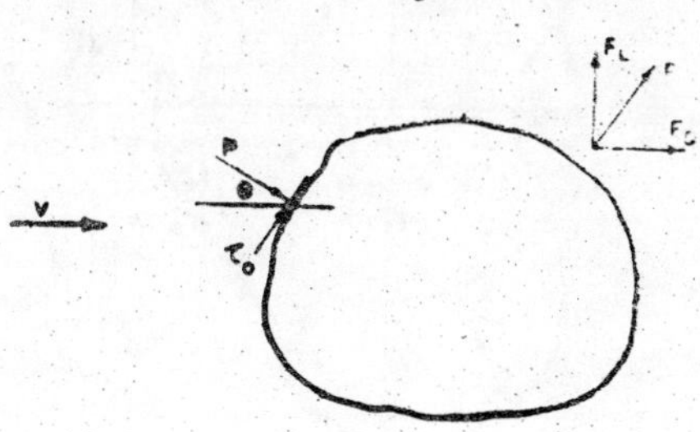

Fig. 2.11.1 Forces acting on an element

If the flow takes past a smooth cylinder at sufficiently high Reynolds number ($R_e > 10^5$), the frictional drag will be too small and can be left out.

$$\therefore F_D = \int_A p \cos\theta \times dA \qquad (2)$$

Dividing both sides by $\rho U^2 A/2$ one gets

$$C_D = F_D / 0.5\, \rho U^2 A = \int_A p \cos\theta \times dA / 0.5\, \rho U^2 A \qquad (3)$$

where C_D is coefficient of drag and A is the projected area of the cylinder normal to the direction of flow and ρ is the mass density of fluid.

4 CALIBRATION OF WIND TUNNEL Reter to Fig. 2.11.3 of the wind tunnel. Let the pressure and velocity of flow at the entrance section of wind tunnel at B be p and U_a respectively. Applying Bernoulli's theorem between A away from entrance and B,

$$\frac{p_a}{w_a} + \frac{U_a^2}{2g} = \frac{p_B}{w_a} + \frac{U^2}{2g} \qquad (4)$$

Since the atmospheric air can be considered to be almost still and hence U_a is approximately zero.

$$\therefore U^2/2 = g\,(p_a/w_a - p_B/w_a) \qquad (5)$$

If an inclined U-tube (slope angle α_o) measures this differential pressure as l_o,

$$\rho_a U_a^2/2 = w_a (S_m/S - 1)(-l_o \sin \alpha_o)$$

—ve sign is used as the pressure at B will be lower than atmospheric pressure.

5 PRESSURE AROUND THE CYLINDER Let l and l_o be the inclined U-tube readings (slope α) corresponding to B and C (a point on the surface of cylinder).

$$(p_C - p_B)/w_a = (S_m/S - 1)(l \sin \alpha - l_o \sin \alpha_o) \qquad (6)$$

Combining (5) and (6)

$$(\Delta p / 0.5\, \rho U^2) = (p_C - p_B)/0.5\, \rho U^2 = (l \sin \alpha - l_o \sin \alpha_o)/l_o \sin \alpha_o \qquad (7)$$

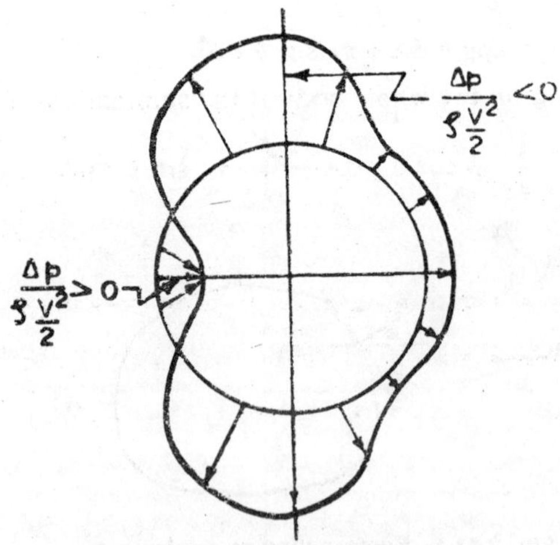

Fig. 2.11.2 Variation of pressure

The values of $\Delta p/\rho\, U^2/2$ are obtained by measurement and plotted along the circumference of the section of cylinder as shown in Fig. 2.11.2. In order to find the value of C_D, the pressure distribution profile is integrated along the circumference of the cylinder of diameter D.

$$C_D = \frac{1}{D}\int_A \frac{\Delta p}{0.5\,\rho U^2}\, ds\ \cos\theta.\ \text{per unit length}$$

Experimental set up

It consists of a wind tunnel having a square shaped test section of about 300 mm size and capable of producing a wind velocity of (about) 10 m/s. A metallic cylinder of 150 mm diameter and 300 mm long is fixed in the wind tunnel with its axis normal to the direction of flow and in such a way that it can be rotated. A protractor is fixed on the outer wall of wind tunnel to measure the rotation of cylinder. A small hole is made on the surface of the cylinder at centre. A small diameter copper tube is attached to it from inside which is then attached to the U-tube manometer. The other end of manometer is left open to atmosphere. The speed of wind tunnel is regulated by means of a rheostate The velocity is measured by noting the pressure at the entrance section (see Fig. 2.11.3).

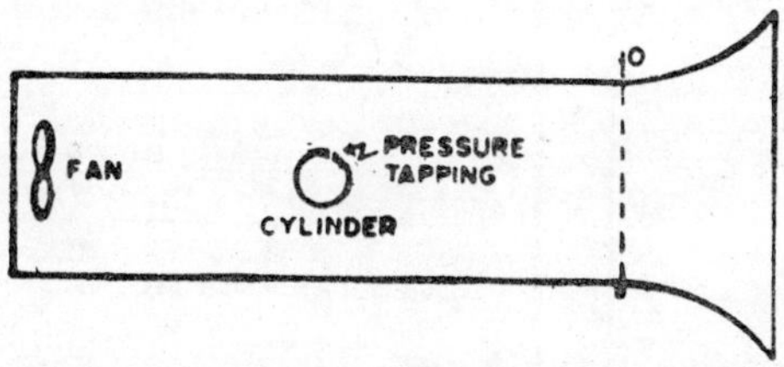

Fig. 2.11.3 Wind tunnel; Bis at the entrance section to the wind tunnel (extreme right section)

Procedure

1. Switch on the exhaust fan of the wind tunnel and adjust the velocity by means of rheostate.
2. Note the pressure at B.
3. Read the protractor and note the reading of second U-tube manometer.
4. Rotate the cylinder by say, 15° and repeat step 3.
5. Repeat the step 4 till the reading all round the cylinder is obtained.
6. Vary the wind velocity by means of rheostate and repeat steps 2 to 5.

Calculations

1. Calculate the velocity of wind stream

$$U = \sqrt{2g\,(S_m/S-1)\; l_o \sin\alpha_o} \qquad \text{(col. 7)}$$

2. Calculate the non-dimensional pressure, $\Delta p/0.5\ \rho U^2/2$

$$\Delta p/0.5/\rho U^2 = (l \sin\alpha - l_o \sin\alpha_o)/l_o \sin\alpha_o$$

Observations

Daimeter of cylinder = m

Length of cylinder = m

Temperature of air = °C

Mass density of air = kg/m^3

Inclination of first manometer α_o =

Inclination of second manometer α =

Set I

Sl. No.	Protractor Reading	l	$l \sin \alpha$	l_o	$l_o \sin \alpha_o$	U	$\Delta p/\rho U^2/2 = \frac{(l \sin \alpha - l_o \sin \alpha_o)}{l_o \sin \alpha_o}$	C_d = (Area of pressure diagram)/D
1	2	3	4	5	6	7	8	9

Set II

Sl. No.	Protractor Reading	l	$l \sin \alpha$	l_o	$l_o \sin \alpha_o$	U	$\frac{\Delta p}{0.5\,\rho U^2} = \frac{l \sin \alpha - l_o \sin \alpha_o}{l_o \sin \alpha_o}$	C_D = (Area of pressure diagram)/D
1	2	3	4	5	6	7	8	9

3. Against a vertical diameter, draw $+\Delta p/0.5\,\rho U^2$ on one side & $-\Delta p/0.5\,\rho U^2$ on the other side for left half of cylinder ($\theta = +90^\circ$ to -90° with respect to wind direction).
4. Similarly draw $\Delta p/0.5\,\rho U^2$ diagram for right hand half of cylinder against the vertical diameter.
5. Find out the net area of pressure diagrams.
6. Compute C_D for the cylinder,

$$C_D = \frac{\text{Net area of pressure diagram}}{\text{Diameter of cylinder}}$$

Presentation of result

1. The pressure diagram is drawn all round the cylinder.
2. The mean value of C_D is determined for the cylinder. Coefficient of drag for the cylinder,

   $C_D =$ ....

Precautions

1. The flow should be uniform and two dimensional.
2. The sign of l and l_o should be noted properly.

Viva-voce

1. What is the importance of the coefficient of drag in the fluid mechanics problems ?
2. What is the effect of variation of Renolds number on the value of C_D?
3. If the length of cylinder is too small so that it behaves like a disc, will the coefficient of drag vary ?
4. Explain why the velocity is maximum and pressure is minimum at the top of the cylinder.

PART 3

Experiments on Open Channel Flow

3.1 TRIANGULAR WEIR OR V-NOTCH

Object

To calibrate a V-notch.

Theory

The discharge Q flowing over notch is given by:

$$Q = C_d \times 8/15 \times \sqrt{2g} \tan \phi/2 \times H^{5/2} = C_d \times K \times H^{5/2}$$

where ϕ is angle of notch, H is head of notch, C_d is coefficient of discharge and

$$K = 8/15 \sqrt{2g} \tan \phi/2$$

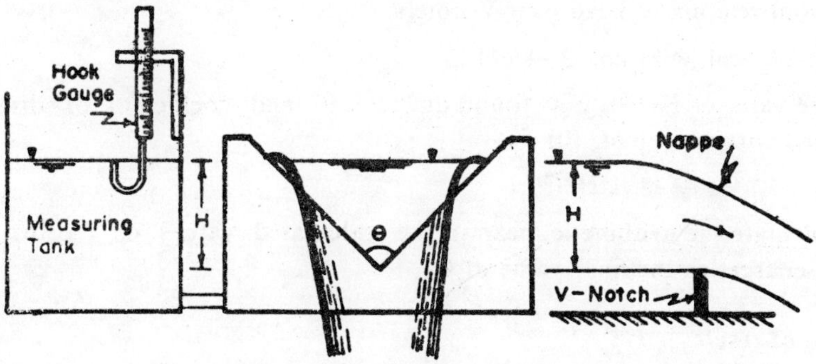

Fig. 2.7 Flow over a V-notch

Experimental set up

The V-notch is fixed in a steel or masonry flume (2 m long, 0.4 m wide and 0.4m deep) to which the water is supplied from a overhead reservoir through a pipe (nearly 225 mm in diameter). The water reaching the notch should be calm and still, i.e. without surface disturbances and without velocity of approach. To achieve this objective, the water is made to pass through baffle walls before it approaches the notch. A hook gauge is used to measure the head of water over the notch and it is placed in an isolated chamber so as to measure a static head free from any fluctuations in water level due to ripples on the water surface. The steel or masonry collecting tank is used to measure the dischage flowing over the notch.

Procedure

1. Initially let the water stand to the vertex of V-notch. Note the gauge reading for Q equal to 0. (col. 2)
2. Close the sluice gates of all the channels carrying water out of collecting tank.

3. Open the delivery valve to allow a constant discharge to flow over V-notch.
4. Wait for some time for water level to become constant in its measuring chamber and note it in col. 3.
5. Now take initial and final readings (col. 4 and 5) of the water levels in the collecting tank below after a lapse of suitable known time t (col. 6).
6. Open the sluice gate of channel No. 1.
7. Repeat the above procedure for seven sets of discharges:

Calculations

1. Calculate the discharge Q as A. (col. 6—col. 5)/t
2. Calculate the constant K of the notch.
3. Determine the head over V-notch as the difference between final reading and initial reading of head over V-notch.

   col. 4 = col. 3 — col. 2

4. The value of $H^{5/2}$ is now found out (col. 9) and coefficient of discharge is then computed (col. 10)

   $$C_d = Q/K\ H^{5/2}$$

5. Calculate the arithmetic mean of the calculated values ofl C_d for different discharges as the final value of C_d.

Presentation of result

Draw a graph between Q and $H^{5/2}$ which should conform to a straight line, passing through a point of Q = 0, H = 0. It shows thereby that the value of C_d is constant at all heads and discharges. Also find the value of C_d from the graph $C_d = Q/K\ H^{5/2}$, i.e. slope of straight line divided by V-notch constant K.

Coefficient for V-notch C_d =

Precautions

1. Note that the final gauge reading over notch is noted when the water level over it has attained a constant value.

2. Throughout one set of readings for a particular discharge, the delivery valve should not be touched. Moreover, if the supply is from a overhead tank, the overflow from the tank should start before experiment is started.

Observations

Angle of V-notch, ϕ =

Angle of collecting tank A = m²

Sl.	Head Reading			Discharge Reading			Discharge	$H^{5/2}$	C_d
No.	Initial m	Final m	H m	Initial m	Final m	Time m	Q m^3/s		
1	2	3	4	5	6	7	8	9	10

Viva-voce

1. What is the use of a V-notch ? What are its advantages over a rectangular notch ?

2. How will you design a V-notch suitable for measuring small discharges very accurately ?

3. Why is the gauge not installed vertically over the notch instead of being installed in a suitable gauge chamber ?

4. What is the use of baffle wall ?

5. How will you ascertain the direction of flow of water over a notch just by seeing the bevelling of its edges ? Justify your answer.

3.2 UNIFORM FLOW

Object

To determine the Manning's and Chezy's coefficient for the given open channel.

Theory

If at any time instant, the velocity and acceleration etc. remain same at all sections of the channel, such flow is termed as 'uniform flow'.

For given uniform flow, the velocity of flow is given by well known resistance equations, e.g. Manning's and Chezy's equations,. Manning's equation,

$$U = \frac{1}{N} R^{2/3} S^{1/2} \quad (1)$$

where U is the mean velocity of flow (= Q/A), R is the hydraulic radius, A is the area of flow corss section,S is the bed slope of the channel and N is Manning's rugosity coefficient which is representative of the roughness of the channel boundary. Chezy's equation

$$U = C \sqrt{RS} \quad (2)$$

where C is known as Chezy's roughness coefficient.

Both these equations are dimensionally nonhomogeneous and hence MKS or SI units should only be used.

Experimental setup

The setup consists of a rectangular (or any other shape) channel to which water is fed at its upstream end. A device for measuring discharge (e.g. V-notch) is provided at its upstream end (or a mesonary tank at its downstream is provided for volumetric measurement).

A point gauge is provided at intermediate section (along the length) of channel. A stop watch is required for discharge measurement.

Procedure

1. Initially let the water stand to the vertex of V-notch. Note the gauge reading for discharge equal to zero.
2. Open the delivery valve to allow a constant discharge to flow over V-notch.
3. Wait for some time for water level to become constant in the measuring chamber and note the water level.

   Alternatively, note the initial and final readings (H_1 & H_2) of the water levels in the collecting tank after a lapse of suitable known time t.
4. Allow the water to flow in open channel at this constant discharge.
5. Take the reading of depth of water at the given cross section at its centre line.

6. Note down the longitudinal slope, width and other geometrical dimensions of the channel.
7. Repeat steps 1 to 6 for seven sets of discharges.

Calculation

1. Calculate the discharge Q by V-notch formula.

$$Q = \frac{8}{15} C_d \sqrt{2g} \tan \theta/2. H^{5/2}$$

Alternatively,

$$Q = A_1 (H_2 - H_1)/t$$

2. Calculate the area A and perimeter P of the channel cross section.
3. Calculate hydraulic radius R (= A/P).
4. The mean velocity of flow U is computed as Q divided by A.
5. Determine Manning's N and Chezy's C using equation 1 and 2

Presentation of result

Mean value of N =

Mean value of C =

Precautions

1. Note that the gauge is in the centre of the corss section of the channel, while the initial and final reading is taken on it.
2. Note that the final gauge reading over the notch (if used) has attained a constant value before it is noted.
3. Throughout one set, the discharge should remain constant.

Viva-voce

1. In Manning's formula, S represents the slope of energy line. How are we justified in taking it as bed slope of the channel for uniform flow computation?
2. Is it possible to have an uniform flow over a horizontal channel?
3. What are different types of depth gauges?

Observations

Angle of V-notch θ =
Coefficient of discharge C_d =
Area of collecting tank A_1 =
Bed width of channel =
Bed slope of channel =
Initial gauge reading over V-notch. =

Sl.No.	Discharge reading				t	Q	Depth reading			A	P	R	U	N	C
	Gauge reading	H	Initial (tank)	Final (tank)			Initial	Final	Depth y						

3.3 PITOT TUBE

Object

1. To determine the velocity distributions in an open channel by Pitot tube and thereby to determine the discharge flowing in it.
2. To determine the energy and momentum correction factors.

Theory

A pitot tube is used to find the velocity of an open stream at any point. A tube bent at 90° at its end and drawn out into a nozzle may be used to find the velocity of stream at a point. The nozzle axis is aligned towards the direction of flow, the water of stream will rise in the tube to a height equal to stagnation pressure head which will project out of the free water surface by the amount of velocity head. Thus the velocity at the tip of the nozzle can be calculated.

$$U = C\sqrt{2g\,(p_s - p)/w} = C\sqrt{2g\,l\,\mathrm{Sin}\,\theta}$$

where 'θ' inclination of the differential pressure gauge to the horizontal, 'l' the deflection of water column in the inclined manometer and 'C' instrument coefficient which is less than unity for most of conventional Pitot tube. It is because the observed value of static pressure p_s is usually less than the actual static pressure. By suitably designing the location of static pressure openings, C can be made unity.

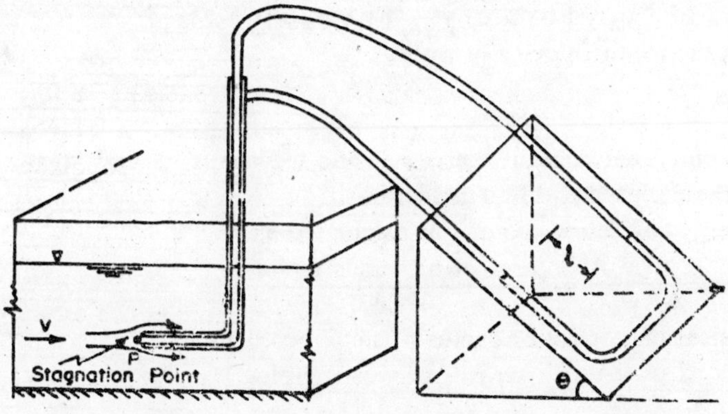

Fig. 3.1 Velocity measurement by a Pitot tube

Experimental set up

It consists of a flume 10 m long, 0.6 m wide and 0.6 m deep which is provided with top rails for the movement of a trolly. A Pitot static tube and a point gauge are mounted on it. The Pitot tube is connected to an inclined U-tube manometer

The inlet pipe to the flume is provided with a regulating valve, and the discharge flowing through pipe is measured by means of a venturimeter.

Procedure

1. Open the sluice valve and allow a certain discharge to flow in the channel. Ensure uniform flow in the channel.
2. Divide whole channel width into four to five sections.
3. Set the Pitot tube in the centre of first section.
4. Take point velocity measurements along the vertical at an interval of 2 cm. or so.
5. Repeat above steps 2 to 4 for all other remaining sections.
6. Note the mercury manometer reading of venturimeter.

Calculations

1. Calculate the discharge flowing into the channel, using given relation for venturimeter.

   $Q = K\sqrt{H}$ (col. 5)

2. Calculate the velocity at different depth

   $u = C\sqrt{2gl \sin \theta}$ (col. 10, 12, 14)

   (*Note:* For streamlined Pitot tube, C is nearly unity.)

3. Plot a graph between velocity (abscissa) and corresponding depth (ordinate) for different sections (col. 9 to 14).
4. Find out the area of this plot (i.e. Σ uy)
5. The discharge through the channel is then computed

   $Q = [\, b_1 \Sigma u_1 y + b_2 \Sigma u_2 y + b_3 \Sigma u_3 y \,]$

6. Find out the mean velocity of water

   $$U = \frac{Q}{BY}$$

7. Prepare plots between u^2 versus y and u^3 versus y and again find the area of these plots for different sections.
8. Then energy correction factor α is calculated as

   $$\alpha = \frac{\Sigma u^3 \, dA}{A U^3} = \frac{\text{area of } u^3 \text{ versus y plot}}{Y U^3}$$

   and momentum correction factor β is computed as

   $$\beta = \frac{\Sigma u^2 \, dA}{A U^2} = \frac{\text{area of } u^2 \text{ versus y plot}}{Y U^2}$$

Presentation of results

1. Compare the discharge determined by Pitot tube and venturimeter.

   Q by Pitot tube =;

   Q by venturimeter =

2. Note down the values of α and β

Observations

Width of flume : m

Depth of water : m

Discharge relation for venturimeter : $Q = K\sqrt{H}$

S. No.	Reading of mercury manometer			Q	Reading of depth gauge			Section I		Section II		Section III	
	Left	Right	H	m^3	Initial	Final	y	l_1	u_1	l_2	u_2	l_3	u_3
1	2	3	4	5	6	7	8	9	10	11	12	13	14

Energy coefficient α =;

Momentum coefficient β=

Precautions

1. Make sure that there is no entrapped air present in the two limbs of manometer connected to Pitot tube.
2. Take readings at all depths.
3. Place the manometer below water surface in the flume.

Viva-voce

1. What do you understand by stagnation pressure ?
2. How will the velocity be affected if the pitot tube is placed at an angle to the direction of flow ?
3. What will happen if the inclined manometer is placed at a higher water surface in the flume ?

3.4 CALIBRATION OF CURRENTMETER

Object

To calibrate the given currentmeter.

Theory

A currentmeter is used to find the velocity at a point in an open stream. It is a mechanical device and essentially consists of a rotating element whose speed of rotation varies with the local velocity of flow. The relation between these two variables must be found by calibration. The (cup type) currentmeter has a ring of conical cups fixed to a vertical spindle mounted in a frame which is lowered into the open stream to the point at which the velocity is to be measured. If the buckets were stationary, the dynamic thrust of the moving stream on the faces concave to it would be greater than on those convex to it. Consequently they begin to revolve and finally reaches a steady speed at which the thrusts are equalised. The indication of speed of rotation is carried out by an electrical contact which is operated by a small battery. The circuit gets closed at every revolution of the spindle and gives a signal which can be heard in the headphone. With the help of stop watch, the speed of revolution of meter N (rev/min), corresponding to the local velocity (velocity at the centre of cups) U in metre per second can be known. A graphical plot of U versus N is the calibration curve of the currentmeter. Calibration constant 'C' of the meter in the calibration formula can be obtained from the graph.

$$U = \frac{N}{C}$$

Experimental set up

It consist of a laboratory flume (10 m × 0.6 m × 0.6 m) to which the water is fed from an overhead tank. The discharge flowing into the channel can be regulated by a sluice valve and is measured by an additional flow measuring device (e g. a V-notch fixed on the upstream end of channel or a venturimeter fitted in the pipe line). The currentmeter and point gauge are mounted on a trolley in such a way that currentmeter can be lowered at different water depths.

Procedure

1. Allow water to stand upto sill of V-notch and take initial gauge reading (col. 2).
2. Allow a constant discharge to flow over V-notch, and then from collecting tank into the channel. (in which the currentmeter is proposed to be lowered.)
3. Let the water level in the tank and water depth in the channel become steady.

Observations

C_d for V-notch =

V-notch constant K (in $Q=K\ H^{5/2}$) =

Bottom width of channel =

S. No.	Gauge Readings								Speed of revolution N			Remarks
	Initial m	Final m	H m	$H^{5/2}$ $m^{5/2}$	Q m^3/s	Depth of flowy m	Area of flow m^2	Mean velocity U m/s	No. of rev. N_1	Time t sec.	No. rev. per min. $N=N_1/60\ t$	
1	2	3	4	5	6	7	8	9	10	11	12	13

4. Observe the constant head over V-notch (col. 3) and water depth in the channel (col. 7)
5. Calculate the mean velocity of flow in the channel (col. 9)

Note: U = Q/B. y where B is the width and y is the water depth of channel.

6. Set the currentmeter so that the centre of its buckets lies at 0.6 times the depth of water below the free surface.

Note: It has been established in practice that in the velocity diagram of steady uniform flow in an open channel, the mean velocity occurs at 0.6 depth below the water surface.

7. Note the speed of revolution (col. 10) corresponding to the mean velocity of flow.
8. Repeat the same with different discharge and thus obtain a set of mean velocity of flow-speed of revolution reading.

Calculations

1. Calcute discharge $Q = K H^{5/2}$ (col. 6)
2. Mean velocity of flow U=Q/A (col. 9)
3. The speed of revolution
   N (r p m) = Number of revolution in t seconds/60 t

Presentation of results

Draw a graph between U (as ordinate) and N (abscissa). It is known as calibration curve of the given currentmeter. Also obtain the calibration constant C

$$C = \frac{N}{U}$$

Precautions

1. Make sure that the gauge reading over notch is constant all over the time the particular set of reading is being taken.
2. The sluice gate in the collecting tank should be raised well above the water level.

Viva-voce

1. Why should we have conical cups in currentmeters ? What will happen if they are replaced by flat blades of the same projected sides ?
2. What is the purpose of tail provided behind cups of the currentmeter? If a currentmeter installed in a position gives constant number of revolutions per minute, what changes will result if the direction of flow is reversed ?
3. Can you strike any similarity between a vane type current meter and the above currentmeter ?

3.5 SURFACE CURVE

Object

To observe the depths and length of M_3 type surface curve and verify the latter by Step Method.

Theory

In a channel, the flow can be uniform or non-uniform. A uniform flow implies a constant discharge and other elements of flow at different sections along the direction of flow. In a non uniform flow, while discharge is constant for continuity, other elements like depths etc. change from section to section. In both cases, the Mannings formula is applicable at a particular section,

$$Q = \frac{1}{N} AR^{2/3} S^{1/2} \quad (1)$$

where Q is the discharge in cumecs, N is Manning's coefficient, A is area of flow, R is hydraulic radius and is equal to $=A/P$; P is perimeter of flow and S is the slope of total energy line and is called friction slope. In case of uniform flow $S = S_o =$ slope of channel bed.

For a given channel section (i.e. channel section and N given), Eq. 1 is a three parameter relationship,

$$f(Q, y, S_o) = 0 \quad (2)$$

for a uniform flow. For a given channel Q and S_o, y is unique and is called uniform or normal depth.

STEP METHOD Water surface profile of non uniform flow in a channel is referred to as "surface curve." The computations of a surface curve can be made by a number of methods. One which is simplest to use is step method, and is described below:

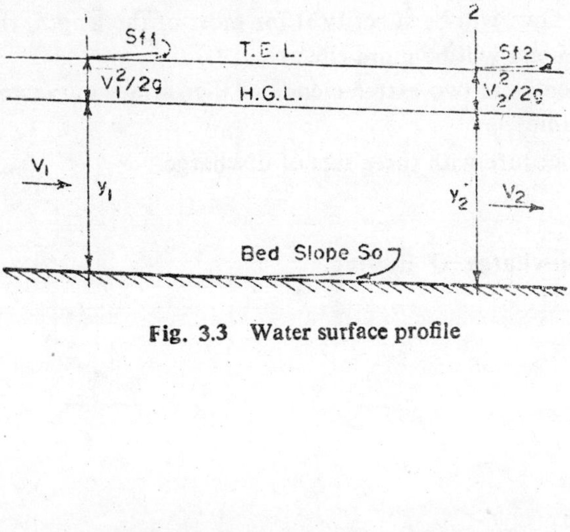

Fig. 3.3 Water surface profile

Consider a channel reach of non uniform flow carrying discharge Q in steady motion. At section 1, the depth of flow is y_1 and velocity U_1 and at downstream section 2, depth of flow is y_2 and velocity U_2. Then distance along the channel between sections 1 and 2

$$\Delta x = \Delta E/(S_o - S_f) \quad (3)$$

$$\text{where } \Delta E = E_1 - E_2 \quad (4)$$

$$E_1 = y_1 + U_1^2/2g \quad (5)$$

$$E_2 = y_2 + U_2^2/2g \quad (6)$$

$$\text{and } \bar{S}_f = \tfrac{1}{2}[S_{f1}+S_{f2}] \quad (7)$$

S_{f1} and S_{f2} can be determined for sections 1 and 2 respectively by using Manning's equation with $S = S_f$. Thus for chosen channel reach with its sections and Manning's N given, Q, y_1, y_2 and S_o observed, other parameters E_1, E_2 or ΔE can be calculated. Finally length of surface curve can be determined from Eq. 3.

Experimental set up

It consists of a flume (preferably tilting) 15 m long in which a sluice gate is mounted at its upstream end and a flap valve at downstream end. The inlet valve of the centrifugal pump regulates the discharge to the flume. A point gauge is used to measure the depth of water in the flume.

Procedure

1. Set the bed of the given tilting bed at a certain inclination and observe it.
2. Start the motor driving the centrifugal pump that supplies discharge to the flume.
3. Ensure a constant discharge by waiting till the mercury deflection in the U-tube manometer (connected to flow measuring device, e.g. venturimeter in the supply pipe) becomes constant and observe the deflection (col. 2).
4. Allow the discharge to flow under the sluice gate and over flap valve at the end of flume.
5. Adjust the sluice gate and flap valve so that a stable flow profile can be obtained which can be identified as follows : a surface curve of M_3 type, a hydraulic jump, a uniform flow reach, and then a drawdown.

Note: Since the flow will be subcritical for most of the length, the flow control at downstream end will be more effective.

6. Observe the depths at two extreme ends of the surface curve and the length between the same.
7. Repeat the procedure with three sets of discharge.

Calculations

1. Calculate the discharge 'Q' flowing.

$$Q = K\sqrt{H}$$

3.5 SURFACE CURVE

Object

To observe the depths and length of M_3 type surface curve and verify the latter by Step Method.

Theory

In a channel, the flow can be uniform or non-uniform. A uniform flow implies a constant discharge and other elements of flow at different sections along the direction of flow. In a non uniform flow, while discharge is constant for continuity, other elements like depths etc. change from section to section. In both cases, the Mannings formula is applicable at a particular section,

$$Q = \frac{1}{N} AR^{2/3} S^{1/2} \tag{1}$$

where Q is the discharge in cumecs, N is Manning's coefficient, A is area of flow, R is hydraulic radius and is equal to $= A/P$; P is perimeter of flow and S is the slope of total energy line and is called friction slope. In case of uniform flow $S = S_o =$ slope of channel bed.

For a given channel section (i.e. channel section and N given), Eq. 1 is a three parameter relationship,

$$f(Q, y, S_o) = 0 \tag{2}$$

for a uniform flow. For a given channel Q and S_o, y is unique and is called uniform or normal depth.

STEP METHOD Water surface profile of non uniform flow in a channel is referred to as "surface curve." The computations of a surface curve can be made by a number of methods. One which is simplest to use is step method, and is described below:

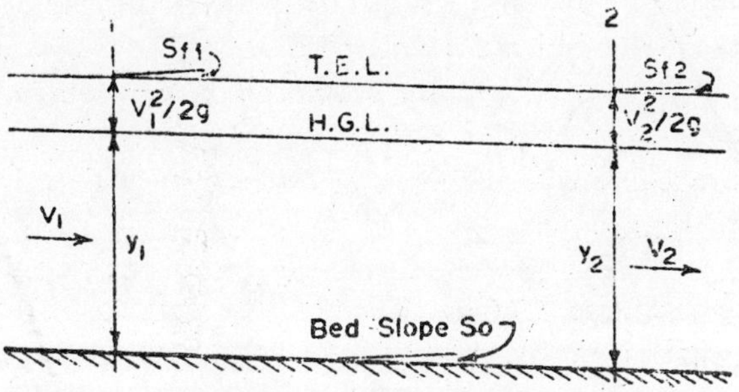

Fig. 3.3 Water surface profile

$\frac{U_2^2}{2g}$	E_2	P_2	R_2	S_{f2}	$E = (E_1 - E_2)$	$\bar{S}_{f2} = {}^1/_2\,(S_{f1} + S_{f2})$	Δx	l_c	% diff.	Remarks
15	16	17	18	19	20	21	22	23	24	25

2. The flow areas A_1, A_2, velocities U_1 and U_2 are then computed (col. 5, 13, 6, 14).
3. The specific energies at sections 1 and 2 are then computed as per Eqns. 5 and 6.
4. Using Eq. 1, S_{f1} and S_{f2} are computed and hence average energy line slope $\bar{S}_f = 1/2\ (S_{f1}+S_{f2})$ is computed. (col. 11, 19 and 21).
5. Finally lengths of surface curve is computed as per Eq. 3.

   $\Delta x = (E_1 - E_2)/(S_o - \bar{S}_f)$
6. Find out the percentage differece in the computed and observed length of curve, i.e. % difference = $100\ (l_o - l_c)/l_o$

Presentation of result

1. Note the computed length of surface curve = metre.
2. Note the observed length of curve = metre.
3. The precentage difference in the observed and computed lengths of curve =

Note :—A small difference in col, 24 ensures satisfactory result.

Precautions

1. Adjust the flap valve and sluice gate so that a stable profile is ensured.
2. Measure water depths accurately.

Viva voce

1. In the following instances, do you expect a uniform or non uniform flow? Flow in a long canal, flow approaching a high dam, flow over a weir spanning the channel width, flow below a sluice.
2. If the sluice gate is kept fixed and flap valve is raised, will it result in shorter or longer surface curve in the above case ?
3. How can the accuracy of computation be increased and value of col. 24 reduced ?

3.6 CRITICAL FLOW METERS

(i) STANDING WAVE FLUME

Object

To determine the discharge coefficient of the given standing wave flume.

Theory

Standing wave flume is a device intended to measure discharge through an open channel. As in a venturimeter, here too, the area of flow is reduced to result in an increase in velocity. This is done by contracting the width and raising the bed thus creating a flumed section. At the flumed section, critical flow will occur and the depth of flow will be critical which, as can be analytically shown, will be equal to two thirds of the height of energy line above the bed at the flumed section. Height of energy line can be taken as height of water surface provided the velocity of water approaching the flume section is negligible. The latter is ensured by providing a honey comb wall upstream. Then energy of water due to increased velocities passing through flumed section should be dissipated before it can cause some harm to the channel bed downstream due to scouring. Therefore, after the flumed section follows a steep slope (to create supercritical depths) and a double row of staggered blocks to dissipate energy (by direct impact) on the channel bed. This arrangment gives rise to creation of hydraulic jump.

The discharge Q through the standing wave flume is given by the equation.

$$Q = K.\, b.\, H^{3/2}$$

where b is the width of flumed section, H is the height of water surface upstream above the hump assuming negligible velocity of approach, and K is the discharge coefficient of the flume.

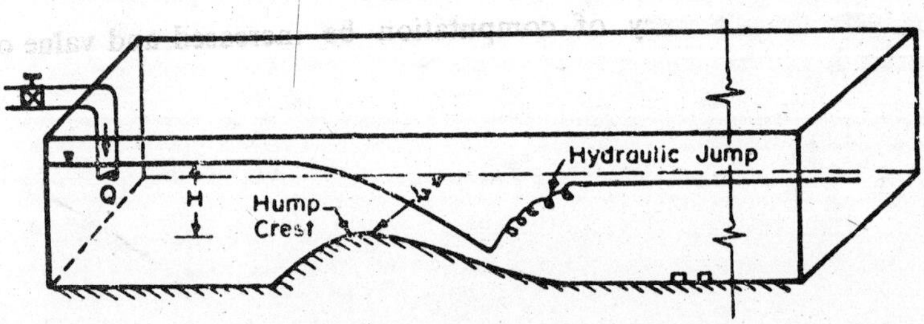

Fig. 3.4 Standing wave flume

Experimental set up

It consists of a standing wave flume along with a point gauge to measure the head of water above hump crest. A tank is required to collect the water and hence to determine the rate of flow. (Fig 3.4)

Observations

Width of flume, b = m

Area of collecting tank, A = m^2

Sl.	Gauge reading for head			$H^{3/2}$	Discharge reading			Q	K
No. m	Initial	final m	Head, H m		Initial m	Final m	Time sec.	m^3/s sec.	
1	2	3	4	5	6	7	8	9	10

Procedure

1. Open the outlet valve slightly and let the water fill the channel upstream up to the hump level. Take the initial gauge reading (col. 2)
2. Let a constant discharge flow in the channel and produce hydraulic jump downstream. Take final gauge reading. (col. 3)
3. Collect the flow in a collecting tank and note the rise in water level(col. 6-7) in a suitable time. (col. 8)
4. Repeat the above procedure for seven sets of discharge.

Calculations

1. The head causing flow is the difference of final and initial gauge readings (col. 4).
2. The discharge Q (col. 9) is determined as
$$Q = \frac{A.\Delta h}{t} = \frac{A\,(\text{col. }7 - \text{col. }6)}{t} \;\frac{m^3}{s}$$
3. The discharge coefficient K (col. 10) is then calculated
$K = Q/H^{3/2} = \text{col. } 9/b \times \text{col. } 5$

Presentation of results

Plot a graph between Q (abscissa) and $H^{3/2}$ ordinate (col. 5) and show that K is constant at all discharges. Find K from the graph. Also calculate the arithmetic mean of its different values for each set of readings (col. 10)

K (graph) =
K (analytical) =

Precautions

1. Note the gauge readings carefully.
2. Before noting the final gauge reading, make sure that hydraulic jump is occurring at the foot of the slope.
3. Use metric system of units as above equation is not dimensionally homogeneous.

Viva voce

1. What is the difference between a venturi flume and a standing wave flume ? What are their uses ?
2. What is the function of baffle wall, staggered blocks, hump, steep slope ?
3. How would you account for approach velocity and friction in discharge formula ?
4. Why do we want a hydraulic jump to occur as the water flows in the flume ?
5. Why does not the water depth at hump fall below the critical depth and how is this critical depth related with total static head in the channel above the hump ?

3.7 CRITICAL FLOW METERS
(2) BROAD - CRESTED WEIR

Object

To calibrate a broad crested weir.

Theory

A control section in an open channel is that section at which a definite relationship exists between the stage and the discharge of flow, e.g. at critical state of flow. The relationship is given as

$$Q^2 T/gA^3 = Z$$

where Q is the discharge flowing in the channel, A is the area of flow and T is the top width of the channel.

When an obstruction is put up across the flow, the critical flow occurs over the obstruction and hence this fact is utilised for measurement of discharge in the open channel.

A weir having a wide crest is known as broad crested weir. The flow pattern over such a weir is shown in figure. The discharge Q passing over it is given by

$$Q = 1.7\, C_d .\, B.\, H_1^{3/2}$$

where B is the width of weir, H_1 is the total head of water above the crest of weir $(= H + U_a^2/2g)$. Neglecting velocity of approach and simplifying.

$$Q = 1.7\, C_d .\, B.\, H^{3/2}$$

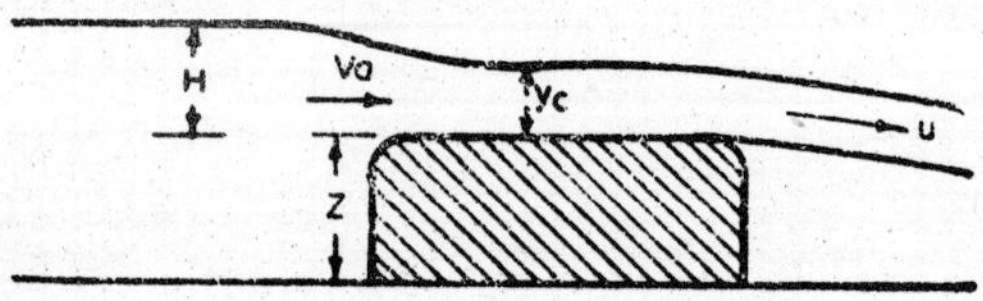

Fig. 3.5 Broad crested weir

Experimental set up

It consist of an open channel or flume 10 m long, 0.5 m wide and 0.5 m deep well plastered on all its sides and bed (Fig. 3.5). The rails are fixed at its top so that point gauges fitted in a trolley could be moved easily. Honey comb walls are provided at upstream so as to minimise surface waves. A broad crested weir 0.4 m wide and 0.15 m high is provided in the channel. The discharge to the channel is controlled by the sluice valve fixed in the supply line. A collecting tank is required to measure the discharge.

Observations

Width of broad crested weirs $B =$ m
Length of broad crested weir $L =$ m
Height of broad crested weir $Z =$ m
Crest level of broad crested weir $H_1 =$ m

Sl. No.	Discharge reading			Q	Final reading of head over weir H_2	Head H over weir	$H^{3/2}$	Cd
	Initial m	Final m	Time sec.	m^3/s	m	m		
1	2	3	4	5	6	7	8	9

Procedure

1. Allow a certain discharge to flow in the flume. Adjust the sluice gate in such a way that steady flow takes place in the flume.
2. Choose a suitable section where the water surface has no curvature at all the discharges.
3. Measure the discharge by volumetric method or otherwise.
4. Repeat the procedure for seven sets of discharges.

Calculations

1. Calculate the discharge $Q = A. \Delta h/t$ (col. 5)
2. The head over broad crested weir $H = H_2 - H_1$ (col. 7)
3. The coefficient of discharge is then determined as
   $C_d = Q/1.7\ BH^{3/2}$

Presentation of results

1. Prepare a graph between Q (abscissa) and $H^{3/2}$ (ordinate) and fit a straight line. Find the slope 'C' of this line; therefrom determine coefficient of discharge.
   $C_d = C/1.7\ B$
2. Also determine the arithmetic mean of the coefficient of discharge.
   C_d (graph) =
   C_d (analytically) =

Precautions

(1) Ensure a steady discharge to flow in the flume.

(2) Select the section properly for the measurement of head, having no curvature of flow. It should be far away from the weir.

Viva-voce

(1) What will be the difference in the flow profiles over a sharp crested weir and a broad crested weir ?

(2) What should be the depth of flow over a broadcrest weir if the depth up-stream is H metre above weir crest ?

(3) Why is it necessary for the flow to be at a critical state over a broad crested weir ?

3.8 CRITICAL FLOW METER
(3) END DEPTH METHOD

Object

To calibrate a free overfall in an open channel.

Theory

The free overfall is a special case of the hydraulic drop. It occurs where the bottom of a flat channel is discontinued. The specific energy curve shows that the section of minimum energy should occur or critical section should occur at the brink. In practice, however, it is not true. The actual situation is that the brink section is the true section of minimum energy but is not the critical section as computed by the principle based on the parallel flow assumption. H. Rouse found that for small slopes the critical depth is about 1.4 times the brink depth or $y_c = 1.4\ y_o$. The discrepancy is due to the fact that the flow at brink is actually curvilinear (for the curvature of flow is pronounced) and the method based on parallel flow is invalid.

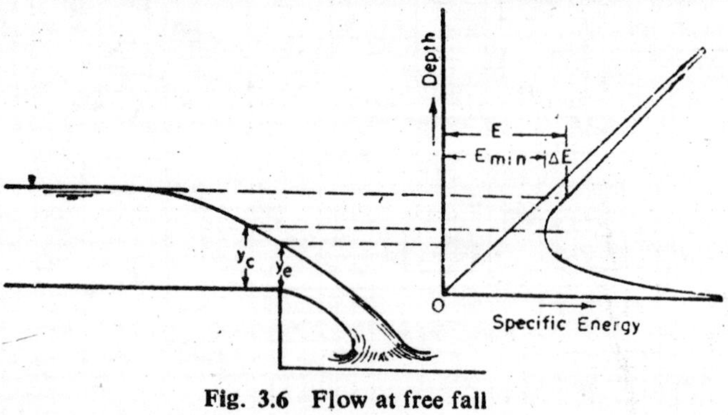

Fig. 3.6 Flow at free fall

Experimental set up

It consists of a glass walled rectangular flume about 4 m long, 30 cm wide and 30 cm deep having honey comb wall at the entrance, Fig. 3.6. A pointer gauge is used to measure the depth of water. A sluice valve is provided near the outlet end of the supply pipe.

Procedure

1. Take the initial reading along the bed of the flume by means of a pointer (depth) gauge. (Starting from the free end or brink and proceeding upstream.)
2. Allow certain discharge to flow in open channel by operating sluice valve and ensure uniform flow in the channel.

Observations

Width of flume = m

S.No.	Manom. deflection m	$Q=C\sqrt{\Delta H}$ m³/sec.	Disch. per unit width $q=Q/B$ m³/sec./m	$y_c=(q^2/g)^{1/3}$	Distance from brink, x	Water depth reading Initial	Final	y	Non-dimensional distances & depths x/y_c	y/y_c
1	2	3	4	5	6	7	8	9	10	11
Run No. 1										
1										
2										
3										
.										
.										
.										
Run No. 2										
1										
2										
3										
.										
.										
.										
Run No. 3										
1										
2										
3										
.										
.										
.										

3. Again take the reading (final) on the water surface along the flume, starting from the brink.

4. Note the deflection of Mercury monometer connected to flow measuring device (Venturimeter) fitted in the supply pipe line.

5. Repeat steps 2 to 4 for three different sets of discharges.

Calculations

1. Calculate discharge $Q = C\sqrt{\Delta H}$ (col. 3)
2. The critical depth of flow

$$y_c = \left[\frac{Q^2/B^2}{g}\right]^{1/3} = (q^2/g)^{1/3} \quad \text{(col. 5)}$$

3. The non dimensional distances and depths are then computed, i.e. x/y_c and y/y_c.

Presentation of result

1. Plot x/y_c (abscissa) versus y/y_c on a simple graph for different discharges.
2. From this smooth plot, find the following values.

   y/y_c (at $x/y_c = 0$), i.e. at the brink =

   x/y_c at $y/y_c = 1$, i.e. distance of critical flow section from the end =

Precautions

1. Take readings at shorter intervals for the water depth near the brink.
2. Ensure a uniform flow in the channel before taking the readings.

Viva-voce

1. What is a control section ? Can the flow at brink be taken as at control section ?
2. The energy at the critical section should be minimum. Then why is the energy at the brink found out to be minimum ? Explain.
3. What type of pressure distribution do you expect at the free end ?

3.7 HYDRAULIC JUMP

Object

1. To study the formation of hydraulic jump.
2. To obtain y_2/y_1, E_L/E_1 and h_j/y_1 versus N_{F1} (Froude's number) for the hydraulic jump.

Theory

By the use of Bernoulli's theorem for steady flow of an ideal incompressible fluid applied to flow in open channels, the total energy contained in unit weight of water along any stream line at any section is give as

$$H = \text{datum head} + \text{depth of water} + \text{velocity head}$$

$$\text{or} \quad H = z + y + U^2/2g$$

It is found that it is specific energy (the energy contained per unit weight of water measured above the channel bed) designated as 'E', which is of great importance in studying the flow phenomenon in open channels.

$$\text{Specific energy } E = y + U^2/2g$$

where y is the depth of flow and U is the mean velocity of flow which is equal to Q/area of flow. For a constant discharge and for given channel section, it can be seen that E is a function of y. A graphical plot of E vs y can either be computed analytically or experimentally obtained. It is known as "specific energy curve." The depth of flow corresponding to least content of specific energy is known as critical depth, designated by 'y_c'. Depth of flow y less than critical depth y_c is referred to as lower stage and corresponding velocity as supercritical velocity. The depth of flow greater than critical is referred to as upper stage and corresponding velocity subcritical velocity

A hydraulic jump occurs in on open channel when the rapid change in the depth of flow is from lower stage to upper stage because of an obstruction in the stream or a change in slope or other channel condition The divergent flow in hydraulic jump is accompained by the formation of extremely turbulent rollers on the sloping surface of hydraulic jump causing a dissipation of large amount of initial energy.

By use of momentum theorem, it can be seen that the depth before the jump, y_1, and after the jump, y_2, is related as

$$y_2/y_1 = \tfrac{1}{2}\left[-1 \pm \sqrt{1+8N_{F1}^2}\right]$$

The energy loss in the hydraulic jump is given as

$$E_L = E_1 - E_2$$

$$\text{or } E_L = (y_1 + U_1^2/2g) - (y_2 + U_2^2/2g)$$

$$\text{or } E_L = (y_2 - y_1)/4y_1\, y_2$$

And initial Frounde's number $N_{F1} = U_1/\sqrt{g\, y_1}$

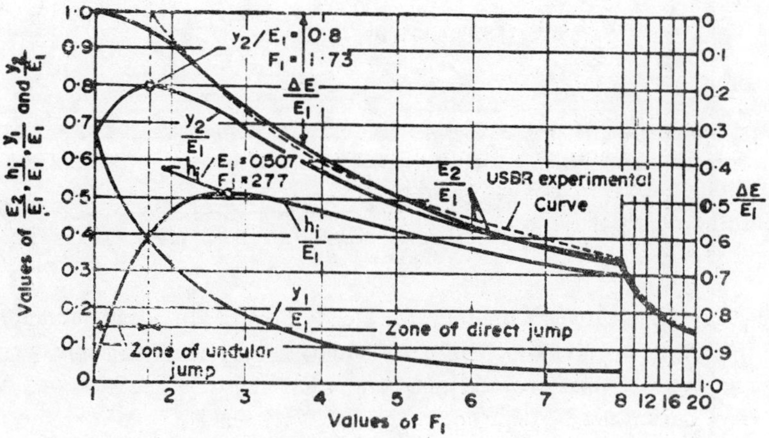

Fig. 3.7 Characteristic curves of hydraulic jump

Experimental set up

It consists of a tilting bed flume to which water is supplied by a pump, Fig. 3.7. The sluice valve in the delivery pipe is used to regulate the discharge The flow measuring device is provided (say, a venturimeter) in the delivery pipe. A sluice gate is fitted on the upstream side of flume and flap valve at its downstream side. The point gauge is an additional instrument needed to measure the depth of flow.

Procedure

1. Set the channel bed of flume at a certain slope, and note the same.
2. Allow some water to flow down the inlet pipe.
3. Adjust the sluice gate opening and downstream (gàte) control such that a stable jump forms in the flume for the given discharge.
4. Note the differential mercury head of 'U'-tube manometer connected to flow measuring device.
5. Measure the depth of flow at a point upstream of the hydraulic jump, i.e. y_1 and depth of flow downstream of hydraulic jump. i.e. y_2 (water surface reading—bed level reading).
6. Repeat steps 2 to 5 for seven different discharges.

Calculations

1. Calculate the discharge Q (col. 4) using proper formula.

$$Q = K\sqrt{\Delta h}$$

2. Specific energy at sections 1 and 2 is then calculated

$$E_1 = y_1 + U_1^2/2g \quad \text{(col. 13)}$$

$$E_2 = y_2 + U_2^2/2g \quad \text{(col. 16)}$$

3. The loss of energy E_L in the hydraulic jump is

$$E_L = E_1 - E_2 \quad \text{(col. 17)}$$

Observations

Width of flume = m

Sl.	Discharge reading			Lower stage			Upper stage			U_1	$U_1^2/2g$	E_1	U_2	$U_2^2/2g$	E_2	E_L/y_1	$\frac{y_2-y_1}{y_1}$	y_2/y_1	N_{F1}
No	Diff. of mercury	Equivalent water head	Q m^3/s	Bed reading	Final	y_1	Bed reading	Final	y_2										
1	2	3	4	5	6	7	8	9	10	11	12	13	14	15	16	17	18	19	20

4. The height of jump is $(y_2 - y_1)$ is nondimensionlised $(y_2 - y_1)/y_1$ (col. 18)
5. Ratio of sequent depths y_2/y_1 is now determined. (col. 19)
6. Initial Froude's number is then entered in (col. 20).

Presentation of results

1. Plot the values of y_2/y_1 (col. 19), $(y_2 - y_1)/y_1$ (col. 18), E_L/y_1 (col. 17), as ordinate and N_{F1} (col. 20) as abscissa on a simple graph.

Viva-voce

1. What is specific force?
2. What is practical application of hydraulic jump?
3. Why does the water surface in the hydraulic jump appear whitish?
4. Differentiate between the energy and momentum principles. Why is momentum principle used for the analysis of hydraulic jump?
5. How is the location of a hydraulic jump controlled by the tail gate?

PART 4

Experiments on Hydraulic Machines

4.1 PELTON WHEEL

Object

1. To obtain characteristic curves (graphs for Q_u, P_u, efficiency vs N_u at full, 4/6 and 2/6 gate openings for given Pelton wheel.
2. To determine specific speed, N_s, for the turbine at design head (of 15 m).

Theory

A Pelton wheel is a tangential impulse turbine, and water strikes its buckets as a free jet (at atmospheric pressure) with a high velocity. Thus wheel starts rotating and develops power. Quantity of water striking the buckets can be controlled by the movement of a streamlined shaped spear working inside the supply pipe nozzle, thus changing the area of flow of the nozzle. Power developed by the turbine shaft is absorbed by the friction offered by the belt under tension wound round the brakedrum.

Normally a turbine is required to work under one set of running conditions. (i.e power, speed, head etc.) for which it is designed. But it may also have to work under different conditions of running. So prior knowledge of how will it function under different possible conditions is of great use, and accordingly the turbine should be tested, observations taken and graphically represented as a family of curves. These curves are known as 'characteristics of the (particular) turbine', Fig. 4.1.

Specific speed of a given turbine, N_s is defined as the speed of a hypothetical turbine, geometrically identical to the given turbine, running at unit head and of such a size as to develop unit power. It is given by

$$N_s = N\sqrt{P}/H^{5/4}$$

where N, P, and H are the speed (in rev. per min), power (in horse power) and head (in metre) simutaneously of the given turbine.

Experimental set up

It consists of a Pelton wheel to which a brake drum is coupled. A belt is wound round the brake drum with its ends tightened by means of screws. The spring balance is provided at both ends. A techometer is connected to the shaft to measure the speed of the wheel. A sluice valve is provided near the outlet end of the supply pipe to feed the water to the nozzle. A centrifugal pump supplies the water under sufficient head to the turbine. A pressure gauge is fitted near the sluice valve to read supply head. The discharge is measured by means of a notch fitted in the discharging channel.

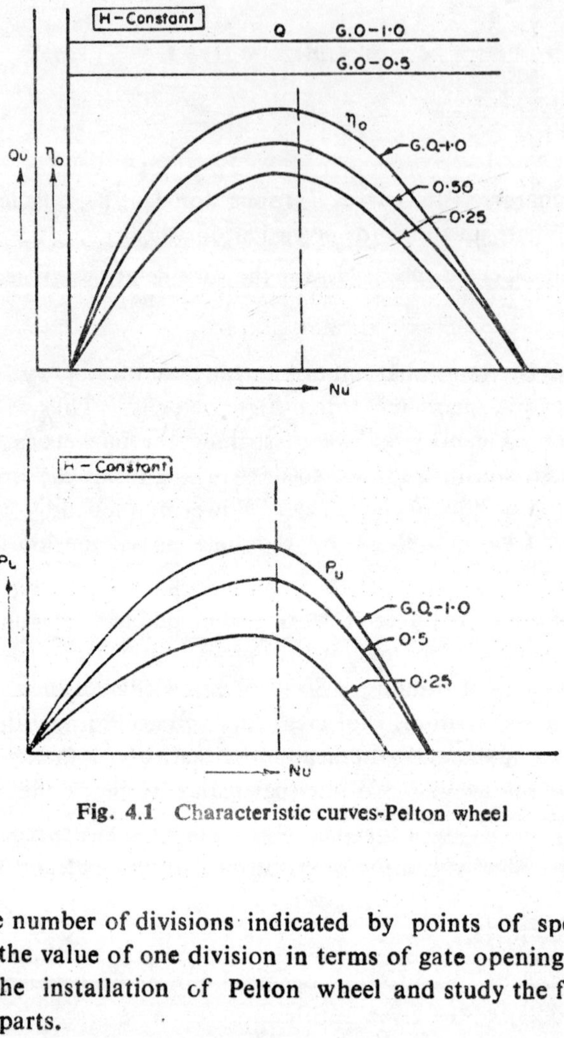

Fig. 4.1 Characteristic curves-Pelton wheel

Procedure

1. Note the number of divisions indicated by points of spear wheel, and as certain the value of one division in terms of gate opening.
2. Study the installation of Pelton wheel and study the function of its component parts.
3. Prime the centrifugal pump and start the electric motor to run the pump. Set the full gate opening by means of spear wheel.
4. Open the delivery valve and set it to a suitable position so as to attain design head of 15 m.
5. Keeping it intact, apply different loads on turbine shaft (by applying weights W_1 and W_2 at the end of belt). Each time note the speed of shaft N also.
6. Note the head over the notch to compute the constant discharge.

Note : So long as gate opening and delivery valve are intact, the discharge Q and net head H will be constant and will be noted once.

7. Repeat steps 4 to 6 for two more gate openings.

Observation

Radius of brake drum = m

Initial gauge reading over notch = m

Sl. No.	Gate Opening	Discharge reading			Spead N rpm.	Power reading			Overall efficiency η_o	P_u	Q_u	N_u
		Final reading m	Head H m	Q m^3/s		Weight W_1 kg	Weight W_2 kg	BHP P				
1	2	3	4	5	6	7	8	9	10	11	12	13

Calculations

1. Calculate the discharge Q passing over the Pelton wheel Q.

   $Q = CH^n$ (col. 5)

Note : Values of 'C'.and' 'n' depend on the type of notch used.

2. The horse power developed by the wheel at the break drum is P

   $P = (W_1 - W_2)\, 2\pi RN/4500$ (col. 9)

3. The input power (in H.P.) is then computed.

   Input power $= WQH/75$ hp

4. Overall efficiency is the ratio of output power to input power to the wheel.

   $\eta_o = P/(w\,Q\,H/75)$ (col. 10)

5. The unit quantities are then calculated.

   $P_u = P/H^{5/2}$

   $Q_u = Q/\sqrt{H}$

   $N_u = N/\sqrt{H}$

Note : It is found very convenient to prepare the characteristic curves in terms of unit quantities.

Presentation of result

1. Plot graphs with N_u as abscissa and Pu, Q_u and efficiency as ordinates separately for different gate openings.
2. From characteristic curves for full gate opening, find N_u and P_u curresponding to maximum efficiency η_o. With known maximum efficiency, discharge and head, calculate P ($= P_u H^{5/2}$).

   (η_o) maximum =

   N_u = r. p. m

   H = metre

   Q = m^3/s

   P = H.P.

3. Determine specific speed to Pelton wheel. $N_s =$ ...... rpm

Precautions

1. Prime the centrifugal pump before starting the motor.
2. Set the gate openings carefully and throughout one gate opening the spear wheel and delivery valve should not be changed.

Viva voce

1. Describe main constructional and principal features of a Pelton wheel. What way is it different from other turbine known to you?
2. Why is it chosen to be tangential flow and not axial flow although performance should have been same?
3. Why is water for Pelton wheel taken from centrifugal pump and not from the overhead tank?
4. What are characteristics of Pelton wheel ? What are their uses?
5. After the nozzle water has atmospheric pressure throughout, then why is a casing provided to the wheel?

4.2 FRANCIS TURBINE

Object

To obtain characteristic curves (a graph between Q_u, P_u, η_o versus N_u at full and 7/13 gate openings) of the given Francis turbine.

Theory

A Francis turbine is an inward flow reaction turbine, in which water enters at high pressure. It discharges water into the tail race through a draft tube. The energy of the flowing water is taken up by the turbine to develop power. Draft tube further assists this cause and water leaving it contains very little energy. Quantity of water fed to the turbine is controlled by movement of movable guide vanes (thus opening out or closing down the passage of water entering the turbine). Actually the quantity of water entering the turbine is under triennial control—control of delivery valve in pipe line leading water to turbine, the guide mechanism as pointed out above, and variation of load on turbine shaft (which varies the shaft speed and hence the centrifugal force opposing the entering water).

Normally a turbine is required to work under one set of running conditions, develop power, speed, head, etc. for which it is designed. But it may have to work under different conditions of running, so, prior knowledge of how will it function under different possible conditions is of great use and accordingly the turbine be tested. Data are obtained and graphically represented by curves. These curve are known as "characteristic curves of the (particular) turbine, Fig. 4.2.

Experimental set up

It consists of a Francis turbine to which water under requisite head is supplied by means of a centrifugal pump. A sluice valve is provided near the outlet end of supply pipe connected to the turbine. A pressure gauge and vacuum gauge measures the pressure head at the inlet and outlet end of the turbine simultaneously. A brake drum is coupled with the Francis turbine to which the load is applied by tightening the belt wrapped round the drum. One spring balance on each end of belt measures the load applied. The speed of running of turbine is measured by means of techno-meter. The turbine discharges in the tail race channel which is measured by means of a V-notch.

Procedure

1. Prime the pump and start the electric motor to run the pump.

2. Open the delivery valve and allow the water to stand just upto the crest of V-notch and then close the valve. Note the initial reading of gauge over the notch.

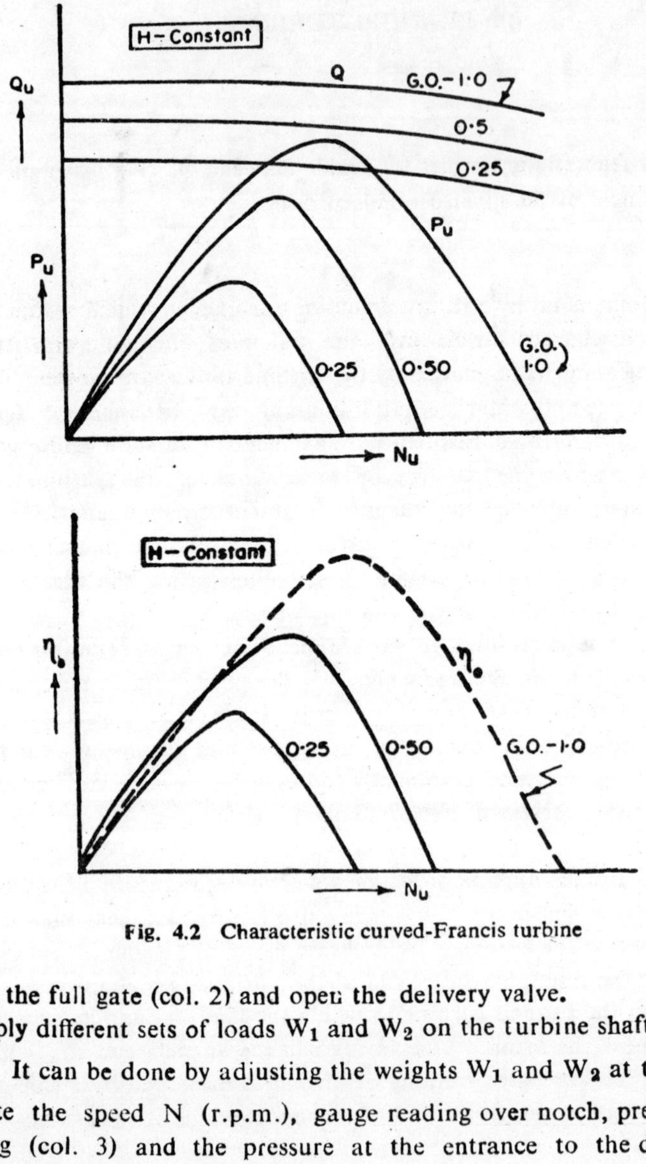

Fig. 4.2 Characteristic curved-Francis turbine

3. Set the full gate (col. 2) and open the delivery valve.
4. Apply different sets of loads W_1 and W_2 on the turbine shaft (col. 10 and 11).

Note : It can be done by adjusting the weights W_1 and W_2 at the ends of belt.

5. Note the speed N (r.p.m.), gauge reading over notch, pressure head at the spiral casing (col. 3) and the pressure at the entrance to the draft tube (col. 4) everytime the loads are changed.
6. Change the gate opening and repeat the procedure.

Calculations

1. The discharge flowing through the turbine is calculated as

$$Q = C H^n \qquad \text{(col. 8)}$$

Observations

Radius of the brake drum R = m

Initial gauge reading over notch, h = m

Sl. No.	Gate opening	Head readings			Discharge reading			N	W_1	W_2	P	η_0	Q_u	P_u	N_u
		Press. gauge	Vacuum gauge	H	Final gauge	h	Q								
		m	m	m	m	m	m^3/s		kg	kg					
1	2	3	4	5	6	7	8	9	10	11	12	13	14	15	16

2. The total net head H under which the turbine is working.

   H = (positive pressure head) – (negative suction head at the entrance to draft tube) (col. 5)

3. The output power is then computed as

   $P = 2\pi RN (W_1 - W_2)/4500$

4. The input power to the turbine is then computed,

   Input power = w QH/75

5. The overall efficiency of the turbine is given as

   η_o = Output power/Input power (col. 13)

6. Finally the unit quantities are calculated (col. 14 to 16)

   $P_u = P/H^{3/2}$; $Q_u = Q/\sqrt{H}$ $N_u = N/\overline{H}$

Presentation of results

In preparing characteristic curves, it is found convenient to determine unit quantities and then plot graphs between Q_u, P_u and efficiency η_o versus N_u, each on separate graph paper for different gate openings. These are known as characteristic curves of the given turbine.

Viva-voce

1. Describe main constructional and principal features of a Francis turbine. What way is it different from Pelton wheel and Kaplan turbine?
2. When the load on the shaft changes, why does discharge entering the turbine change and how?
3. What is the purpose of a draft tube? Why should it be divergent? Why can't we have a greater angle of divergence? Why can't we extend the length of the tube more and more? What will happen if some air leaks in the entrance of the draft tube?
4. What are characteristic curves of a turbine and what are their uses?
5. What are unit power and unit speed?
6. What is specific speed? How does the knowledge of the value of specific speed for a turbine help in selection of the type of turbine? What do you expect its value for this turbine?
7. What is the fuction of spiral casing of the Francis turbine?

Observations

Turbine speed	=	1000 rpm
Power factor	=	0.8
Efficiency of alternator	=	50%
Initial gauge reading	=	m
Full load on the turbine	=	15 HP

Sl. No.	Load	Line current	Line voltage	Head reading			Discharge reading							
				Pressure gauge m	Vacuum gauge m	H m	Final m	h m	Q m^3/sec	P watts	P_1 m HP	% of full load	% η_o	Remarks
1	2	3	4	5	6	7	8	9	10	11	12	13	14	15

5. Throw more load on the alternator and turbine and repeat steps 3 and 4.
6. Repeat till all the loads are put on the alternator or it is not possible to maintain a constant speed at 1000 rmp.

Calculations

1. The net water head on the turbine is H

   H = (positive pressure head) — (negative vacuum head) (col. 7)
2. The discharge fed to the turbine Q

   $Q = CH^n$

Here the values of C and n depend on the types of notch used.

3. The power required by the alternator is P

   $P = \sqrt{3}\ V\ I \cos\theta$ watts (col. 11)
4. The power developed by the turbine is P_1

   $P_1 = P/735\ \eta_A$ (col 12)
5. The percentage of full load is then found out as the ratio of part load to full load. (col. 13)
6. The power input to the turbine is given by P_2

   $P_2 = W\ Q\ H/75$
7. The over all efficiency η_o of the turbine is calculate as

   $= P_1/P_2 \times 100$ (col. 14)

Presentation of results

Draw a graph between % of full load as abscissa and percentage of efficiency as ordinate. This is known as constant speed characteristic curve.

Precautions

1. At the end of experiment, close the delivery valve before removing the load.
2. Initial reading of head over the notch should be taken carefully.

4.4 CENTRIFUGAL PUMP

Object

1. To obtain characteristic curves (graphical plot of head, power and efficiency versus discharges (for the given centrifugal pump running at constant speed.
2. To determine its specific speed.

Theory

The action of a centrifugal pump is that of a reversed Francis turbine with special arrangements to increase its efficiency. Water enters the pump at its centre through a suction pipe, at the end of which is provided a screen (or strainer) and a foot-valve. Screen prevents any debris to enter the suction pipe lest it may clog the blade passage. Foot valve establishes unidirectional flow i.e. into the suction pipe only. This is useful in priming of the pump. Pump is coupled to an electric motor which usually is the driving unit, hence the performance test of the centrifugal pump is carried out at the design speed only, So keeping the speed constant, the pump may be made to discharge different quantities at different heads. Corresponding power input to pump shaft may be known and efficiency can be calculated. The shaft H.P., input P, Head H and efficiency η_o can be plotted against the discharge Q litre/sec. Resulting curves are known as characteristic curves of the (particular) centrifugal pump.

The specific speed of a centrifugal pump is defined as the speed of a hypothetical pump, geometricaliy similar to the given one and of such a size, that will deliver one litre/sec. under one metre head. If the given pump delivers Q litre/sec. under a head of H metre at maximum efficiency (design condition) while running at N revoluti on per minute, then

$$\text{Specific speed, } N_s = N\sqrt{Q}/H^{3/4}$$

Experimental set up

It consists of a 30 H.P. two stage centrifugal pump driven by a 3 phase electric motor. The pump delivers water to a overhead tank through a delivery valve and a delivery pipe. An orificemeter is fitted in between the delivery line to measure the discharge (method has been described earlier). A vacuum guage is provided at the entrance to the spiral casing and a pressure guage at the exit of it to determine the total static head developed by the pump. The power input to the electric motor is measured by means of an energy meter, fitted on the electric line to the motor.

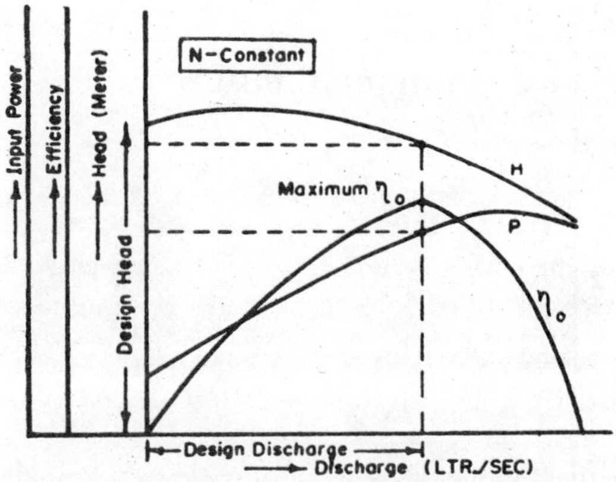

Fig. 4.4 Characteristic curves-Centrifugal pump

Procedure

1. Study the centrifugal pump installations and also study the functions of its component parts.
2. Close the delivery valve and prime the centrifugal pump

Note: The water is filled in the pump casing, suction pipe and the delivery pipe upto the delivery valve.

3. Start the electric motor to run the pump to its full speed.
4. Wait for few minutes and then open the delivery valve.
5. Note the reading of wattmeter (say K Kilowatts), readings of pressure gauges connected to orificemeter (say p_1 and p_2) and readings of vacuum gauge (p_s) pressure gauge (p_d).
6. Repeat the steps 1 to 5 for various settings of delivery valve and hence various discharges.

lations

1. Calculate the discharge supplied by the pump.

$$Q = 10^3 \, C \sqrt{\Delta h} \text{ lit/sec.} \qquad \text{(col. 5)}$$

ote: The meter constant C ($= C_d \, a \sqrt{2g}$) is computed for the given eter.

The total head produced by the pump is given by

= (pressure gauge reading) — (negative vacuum gauge reading) ± (adjust-he location of pressure gauge/vacuum gauge).

If pressure and vacuum gauges are fixed h_1 and h_2 metres above centre trifugal pump, then.

$$_d - (-p_s) + (h_1 + h_2) \qquad \text{(col. 8)}$$

shaft horse power input to the pump is then calculated

Observations

Speed of electric motor N = rpm;

Efficiency of electric motor η_M = %

Sl. No.	Discharge reading				Head readings			Wattmeter reading	Input SHP	Efficiency
	p_1	p_2	Δh	Q	Vacuum gauge	Pr. gauge	Head	'K		
	kg/cm²	kg/cm²	m	lit/sec.	m	m	m	KW	mHP	%
1	2	3	4	5	6	7	8	9	10	11

$P = \sqrt{3}\ K\eta_M / 0.736$

Here η_M is the efficiency of electric motor (col. 10)

4. The power output of centrifugal pump is calculated as

output $= w\,Q\,H/75$

5. The overall efficiency is then finally calculated

$$\eta_0 = \frac{\text{Output}}{\text{Input}} = \frac{w\,Q\,H}{75} \qquad \text{(col. 11)}$$

Presentation of results

1. Draw a graphical plot of H (col. 8), efficiency η_0 (col. 11), input power P (col. 10) as ordinate and Q lit/sec. (col. 5) as abscissa. The resulting curves are known as characteristic curves of this centrifugal pump.

2. Draw a vertical line through the point of maximum efficiency in the above graph. Thus find the head H (metre) and discharge Q (lit/sec.) corresponding to maximum efficiency. Now calculate the specific speed of the pump N_s rpm

$N_s = $.........rpm

Precautions

1. Before starting the electric motor make sure that the priming of the pump has been computed.

2. During priming and upto the time the pump has run to its full speed the delivery valve should be kept closed. Before stopping the electric motor, the delivery valve should be closed.

Viva-voce

1. What is a pump? What is the principle of working of a centrifugal pump? What is the difference in principle of action of the centrifugal pump and reciprocating pump?

2. What is priming? Why and how is it done before starting the pump?

3. What is the function of foot valve and a strainer (or screen)? How is foot valve helpful in priming?

4. On what factors does the reading of vacuum gauge depend? If the speed of the pump shaft be increased, how will the vacum gauge reading change?

5. Why do we keep the delivery valve closed during priming and upto the time the pump develops its full speed? Also why should delivery valve be closed before stopping the pump?

4.5 HYDRAULIC RAM

Object

To obtain the characteristic curves (graphical plot of efficiency, waste water per minute and useful water per minute versus number of beats per minute) for the given hydraulic ram.

Theory

Normally the pumps derive their energy from external source (i.e. electric motor etc.) in the form of mechanical energy. But a hydraulic ram is a hydraulically driven pump which utilises the dynamic pressure of water flowing under a low head to lift a portion of this water to a higher head. If large quantity of water is available at low pressure, a part of it can be converted into water at higher pressure, the remaining quantity of water going to waste. The hydram can be employed if some natural source of water carrying abundant supply is available, then, only such a wastage of water can be afforded. There is no external power used in working this machine. Therefore it is very cheap.

To understand the principle on which its working is based, assume a constant diameter pipe of length l provided with a valve at its end in which water flows with a velocity U. Now if flowing water is brought to rest by closing the valve (time of closure t), then the inertia of water will give rise to high inertia pressure in the pipe which is given by:

$$\text{Inertia head } h_i = p_i/w = l/g \times U/t$$

So, during the closure of the valve, a retarding head ($-h_d$) will act upon the water column and a reverse rising hydraulic gradient will be set up in the pipe. It may be observed in the above equation that if the valve closes instaneously, the developed pressure may be very high (not infinite, since the above equation is based on certain simplified assumptions).

Taking losses in the pipe line and valves into account, D Aubuissen efficiency η_A of the ram is given by:

$$\eta_A = \frac{W_2 . h_d}{(W_1+W_2)h_s}$$

Here W_1 is the weight of water rejected to waste per minute, W_2 is the weight of water lifted up per minute, h_d is the head of water delivered by ram and h_s is the head of water supplied to ram

It consists of a supply pipe connected at its upper end to the supply reservoir and its lower end to a valve box. The valve box has two automatic valves i.e. a waste valve or impulse valve opening downwards and a delivery valve opening upwards.

An air vessel is fixed above the valve box. The delivery valve is fixed at the foot of air vessel, Fig. 4.5

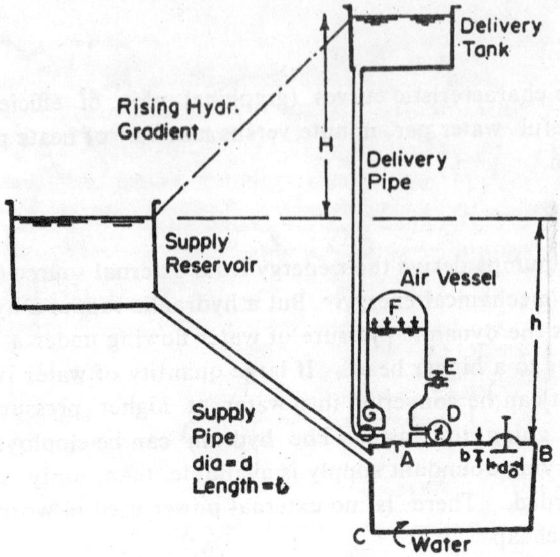

Fig. 4.5 Hydraulic ram

Procedure

1. Firstly open the inlet valve to supply tank and make sure that water is flowing out from overflow pipe of the supply tank.
2. Open inlet valve connected to supply pipe near the hydram.
3. Note the reading of pressure gauge in metre of water (for the particular setting of the lift of the waste valve).
4. Measure discharge of delivery pipe (useful water W_2).
5. Collect the water flowing out through the waste valve in time t_1 and hence determine waste water W_1 flowing out per minute.
6. Count the number of beats of the waste valve occurring in t_3 minutes.
7. Repeat the steps 3 to 6 for various settings of the lift of waste valve.

Calculations

1. The weight of water going waste per minute W_1 is equal to the weight of water collected in time t_1 divided by time t_1. (col. 5)
2. The weight of useful water W_2 is equal to the weight of water supplied in time t_2 divided by time t_2. (col. 8)
3. The ratio of the total number of beats to the time t_3 is computed $N = N_1/t_3$.
4. The efficiency of ram,

$$\eta_A = W_2 h_d/(W_1+W_2) h_s \quad \text{(col. 10)}$$

Observations

Pressure gauge reading h_1 =

Difference of levels between pressure gauges and waste valve h_2 = 0.15 m

Delivery head h_1+h_2

Sl. No.	No. of beats per/sec.	Waste water discharge reading W_1			Useful water discharge reading W_2			$W_1 + W_2$	Efficiency η_A
		Weight of water (g)	Time min	W_1 g/min.	Weight of water (g)	Time min.	W_2 g/min.	(g)	
1	2	3	4	5	6	7	8	9	10

Presentation of results

Draw a graphical plot between useful water flowing per minute (col. 8), waste water flowing per minute (col. 5) and efficiency (col. 10) versus number of beats per minute (col. 2). The resulting curves are known as characteristic curves of the given hydraulic ram.

Precautions

1. Before starting the hydram make sure that the supply tank overflow pipe is running.

2. Lift of the waste valve should be gradual and equally increased (from 0 to 180°) for different readings

3. For each (particular) setting of lift of waste valve the locknut of waste valve must be tightened befor taking the reading for that (particular) setting.

4. It is more important to ensure that air vessel is properly charged with air When it is so, the air inlet valve in the neck of hydram with a pinhole in it should eject a fine jet of water at each stroke of the hydram with a hissing sound and draw in a little air to keep the air vessel charged

5. If hissing ceases, the air vessel is short of air and should be charged.

Viva voce

1. What is the working principle of hydram? Where is it appropriate to use?

2. What form of energy is used to supply the water?

3. If the air vessel is not fully charged with air, will it supply the water to supply reservoir or not?

BIBLIOGRAPHY

1. Alam Singh & Gupta, V.P., " Fluid Mechanics and Machines ", CBS Publishers & Distributors Pvt. Ltd. (Under Publication)
2. Addison, H., "A Treatise of Applied Hydraulics", Chapman and Hall Ltd., 1956.
3. Chow, V.T. "Open Channel Flow". McGraw Hill Book Company, New York, 1959.
4. Garde, R.J., "Fluid Mechanics", Nem Chandra and Bros., Roorkee (India), 1983.
5. Pao. H.F., "Fluid Mechanics", John Wiley and Sons, New York, 1961.
6. Rouse, H. "Elementary Fluid Mechanics", John Wiley and Sons, New York, 1946.
7. Rouse, H., "A Engineering Hydraulics", John and Wiley and Sons, New York, 1950.
8. Streeter, V.L., "Fluid Mechanics", McGraw Hill Book Company, New York, 1948.
9. Venard, John. K., "Elementary Fluid Mechanics", John Wiley and Sons, New York, 1962.

APPENDIX I

Properties of Some Common Fluids at 20°C and Atmospheric pressure

Sl. No.	Fluid	Mass density MKS Unit mslug/m³	Mass density SI Unit kgm/m³	Specific Weight MKS Unit kg(f)/m³	Specific Weight SI Unit N/m³	Kinematic viscosity MKS and SI Units m²/s
1.	Water	102	1000	1000	9810	1×10^{-6}
2.	Carbon tetra chloride	162.5	1594	1594	15632	604×10^{-7}
3.	Alcohol	80.4	789	789	7738	1.5×10^{-6}
4.	Glycerin	129.3	1268	1260	12356	6.3×10^{-4}
5.	Mercury	1380	13533	13550	132880	1.16×10^{-7}
6.	Benzene	88	860	860	8434	7.43×10^{-7}
7.	Air	0.126	1.236	1.208	11.85	1.5×10^{-5}

APPENDIX II

Properties of Air at Atmospheric Pressure

Temperature °F	Temperature °C	Mass density mslug/m³	Specific wt, w kg(f)/m³	Kinematic viscosity m²/s
0	—17.68	0.141	1.571×10^{-6}	1.171×10^{-5}
20	—6.67	0.135	1.708×10^{-6}	1.236×10^{-5}
40	4.44	0.130	1.767×10^{-6}	1.356×10^{-5}
60	15.56	0.125	1.825×10^{-6}	1.468×10^{-5}
80	26.67	0.120	1.879×10^{-6}	1.570×10^{-5}
100	37.78	0.115	1.933×10^{-6}	1.672×10^{-5}
150	65.56	0.107	2.064×10^{-6}	1.923×10^{-5}
200	93.33	0.098	2.191×10^{-6}	2.230×10^{-5}

APPENDIX III

Some of the Important Conversion Factors

Length:

1 cm = 0.394 in
1 m = 3.28 ft
1 km = 0.621 mile

Area:

1 cm^2 = 0.155 in^2
1 m^2 = 10.764 ft^2
1 hectare = 10^4 m^2 = 2.471 acre
1 acre = 43560 ft^2 = 0.4047 hectare

Volume:

1 cm^3 = 0.061 in^3
1 m^3 = 35.315 ft^3
1 litre = 0.0353 ft^3
1 litre = 0.220 imp. gallon
1 litre = 9.264 U.S. gallon

Mass:

1 mslug = 0.672 slug
1 kg (m) = 2.205 lb (m)
1 kg (m) = 0.0865 slug
1 slug = 32.2 lb (m)
1 mslug = 9.81 kg (m)

Force:

1 kg (f) = 2.205 $lb_{(f)}$
1 N = 10^5 dynes
1 N = 0.1020 kg(f)
1 tonne = 1000 kg = 0.984

Mass density :

1 mslug/m^3 = 0.019 slug/m^3
1 gm/cm^3 = 102 muslug/m^3
1 gm/cm^3 = 62.4 lb (m)/ft^3

Weight density:

1 kg (f)/m^3 = 0.0624 lb(f)/ft^3
1 kg (f)/m^3 = 9.81 N/m^3

Pressure:

1 kg (f)/cm^2 = 14.22 lb (f)/in^2
1 kg (f)/m^2 = 0.205 lb (f)/ft^2
1 N/m^2 0.1020 kg (f)/m^2

Viscosity:

1 Poise = 0.1 N-S/m^2
1 N-S/m^2 = 1020 kg (f) sec/m^2
1 kg (f)-sec/m^2 = 0.205 lb (f)-sec/ft^2

Kinematic viscosity:

1 stoke = 1 cm^2/sec = 0.010764 ft^2/sec

Work, Energy, Power:

1 J = 1 N-m = 10^7 ergs
1 J = 0.1020 kg(f)m
1 Kwh = 3.60×10^6J = 0.367×10^6 kg(f)m
1 $kg_{(f)}$m = 7.233 ft lb(f)
1 m.h.p. = 75 $kg_{(f)}$ — m/sec = 0.986 h.p. = 735.5 W
1 h.p. = 550 ft-$lb_{(f)}$ /sec = 746 W = 76.04 $kg_{(f)}$ — m/sec
1 W = 1 J/s = 1 N-m/s = 10^7 ergs/sec